Process First
The evolution of the Business Process Expert

A Collaborative Book by:

**Marco ten Vaanholt and
the SAP Business Process Expert community**

Evolved
Technologist
Press

Process First
The Evolution of the Business Process Expert

A collaborative book by
Marco ten Vaanholt and the SAP Business Process Expert community

Copyright© 2008 by Evolved Media Network, LLC
All rights reserved. Printed in the United States of America.
No part of this book may be reproduced or transmitted in any form or by any means, electronic or mechanical, including photocopying, recording, or by any information storage and retrieval system, without permission in writing from the publisher.

Published by Evolved Technologist Press, an imprint of Evolved Media, 242 West 30th Street, Suite 801, New York, New York 10001

This book may be purchased for educational, business, or sales promotional use. For more information contact:

Evolved Technologist Press
(646) 827-2196
info@EvolvedTechnologist.com
www.EvolvedTechnologist.com

Editor/Analyst: Dan Woods, Deb Cameron
Copyeditor: Deb Cameron
Production Editor: Deb Gabriel
Cover and Design: Deb Gabriel
Cover Mural: Nancy Marguiles
First Edition: September 2008

While every precaution has been taken in the preparation of this book, the publisher and authors assume no responsibility for errors or omissions or for damages resulting from the use of the information contained herein.

ISBN: 978-0-9789218-7-3; 0-9789218-7-9

Contents

Foreword by Zia Yusuf vii

Preface xv

1 The Business Process Perspective 1

2 The Solution Creation Process 15

3 Completing the Definition of the Business Process
  Expert Role 31

4 The Business Process Expert Community 59

5 Organizational Change Management and the Business
  Process Expert 75

6 The Technology Environment for the Business
  Process Expert 95

7 Patterns of Success 109

Afterword 115

Foreword

by Zia Yusuf, Executive Vice President, Global Ecosystem & Partner Group, SAP AG

The creation of a book on the business process expert role by the SAP Business Process Expert (BPX) community is an excellent illustration of SAP's enduring dedication to collaboration and innovation. Both the business process expert role, and the community that surrounds it, show how, in every way possible, SAP is attempting to lead the creation of collaborative structures tailored to the needs of each specific player in the customer and partner universe.

As the pace of business change accelerates and businesses become increasingly connected, collaboration is a key resource, and business networks provide the new source of competitive advantage. Business process experts play an important role in this transformation, since a deep understanding of process and of the orchestration of dynamic business networks is key to delivering higher shared customer value, speed of innovation, cost benefits—and competitive differentiation.

With a deep history of business process expertise, it was only natural for SAP to play an early role in acknowledging the emergence of the business process expert. In 2006, as part of our ecosystem strategy, SAP formed the Business Process Expert community (bpx.sap.com)—one of our Communities of Innovation. The Business Process Expert community has quickly become the leading global business process community and examines the subject of business process management from the perspective of a comprehensive set of industries and horizontal areas.

Within the community, industry experts, business analysts, implementation and application consultants, process developers, enterprise architects, and many others engage in moderated forums, wikis, expert blogs, and other collaborative tools to drive process innovation and solution design. The community collaborates across company boundaries, shares ideas, develops and refines business processes, and leverages the benefits of service-oriented architecture (SOA). Through collaboration, best-practice sharing, and collective learning, the community bridges the gap between business and IT and enables business process platform adoption.

Orchestrating a Customer-Focused Ecosystem

To foster collaboration among customers, partners, and individuals, SAP has embraced an ecosystem strategy that brings together award-winning solutions, comprehensive services, and the power of collaboration among diverse companies and individuals, to give customers the power to operate more effectively.

Every day, our customers are tasked to deliver breakthrough results through accelerated innovation and improved ROI. Service-oriented architecture transforms the way software is developed and implemented. Faced with competing investment alternatives and constant pressure for IT to keep pace with shifting business demands, customers are rethinking success and what it takes to be agile, integrated, and flexible to respond to change and complex pressures—to do more, and deliver more effectively.

SAP believes that maximizing success requires an ecosystem approach that puts the customer in the center. Dealing with market challenges and pressure to innovate at speeds that often exceed business capacity, SAP customers trust they can turn to the SAP ecosystem to collaborate more effectively, compensate for weak areas, decrease time to market, improve performance, and access information more effectively.

> **The Rising Importance of IT and the Business Process Expert in Enabling Collaboration**
>
> According to a recent survey conducted by BusinessWeek Research Services and commissioned by SAP AG ("Getting Serious About Collaboration: How Companies Are Transforming Their Business Networks," June 2008), C-level executives are increasingly turning to collaboration as a way to win new markets and address quickly evolving customer needs. The executives surveyed plan to expand their collaboration efforts even further over the next three years, and emphasized the importance of information technology (IT) and its role in facilitating integration to support their companies' business goals of increased levels of collaboration with customers, partners, and suppliers.
>
> While the findings of this survey confirm that collaboration is being increasingly recognized as a pathway to growth, innovation, and competitive differentiation, only half of the C-level executives responding to the survey are confident that their IT infrastructures will be able to support their collaboration strategies during the next three years. This result underscores the crucial role IT plays in facilitating collaboration and enabling business transformation within the enterprise. While CEOs are embracing the concept of developing customer-centric business models by optimizing the company's network of employees, suppliers, customers, partners, and distributors, IT—and the business process expert—need to play a strategic role to make it all work.

The SAP ecosystem includes:
- Trusted, targeted partner solutions and services
- Unique communities of innovation
- Flexible business process platforms for collaboration

Trusted, Targeted Partner Solutions and Services

As a valuable advisor to customers, SAP provides trusted, targeted partner solutions and services through many initiatives and programs.

To meet this goal, we work closely with leading software partners, technology partners, and service partners. And, we draw on our more than 30 years of experience in creating world-class, business-process solutions, along with industry specific solution maps, to make sure customers have access to the right partners, with the right solutions, at the right time.

Unique Communities of Innovation

Every SAP customer has the opportunity to be a co-innovator. SAP ecosystem communities encompass customers, partners, and individuals who share ideas, technical and business expertise, and experiences to tackle critical business needs. SAP's community-powered ecosystem provides customers with a task- and role-based approach to quickly and easily engage with the wealth of resources available.

Consistent Foundation for Collaboration with the Business Process Platform

All SAP solutions share key components that address common, cross-industry business processes in a unique business process platform. All SAP ecosystem members can collaborate to create extensible, compatible, and fully integrated business solutions that meet customer and partner needs. A common approach and language enables the SAP ecosystem to create valuable service definitions, co-innovate across industries, and enable participation and leadership in industry standards groups.

Through our ecosystem of partners and customers, we are helping to accelerate innovation and improve return on investment for our customers. The SAP ecosystem brings together diverse relationships, resources, and communities to help create the next generation of technology solutions in concert with our own development efforts. Our approach takes us well beyond the partnering model followed by many firms in our industry. The ecosystem is fueled by the collaborative process—connecting SAP with partners, customers, and individuals to achieve common goals. And there are many implications of this collaborative approach along with our focus on maximizing customer value. Let's examine three of those implications.

- The first implication is that **the ecosystem must have the proper business context**. Customers expect that solutions will be appropriate for their industry and meet their business process needs. For that reason, the SAP ecosystem has a very strong industry focus where customers and partners can shape industry solutions and product roadmaps. For example, through the ecosystem we provide Industry Value Networks where customers and partners work together to develop leading industry solutions. We also have, at the first release of this book, more than 400,000 members in our Business Process Expert community who share business process best practices and thousands of partners who help deploy SAP software in specific industries.
- The second implication is that **the ecosystem resources need to be quickly accessible and appropriate to the needs of the specific person or company.** This is accomplished through our Communities of Innovation, which provide ecosystem stakeholders with the ability to leverage the power of community to get what they need, when they need it. SAP's Communities of Innovation provide a role- and task-based approach to engage with the wealth of resources available through the ecosystem. Developers can engage with more than 1.2 million other developers in the SAP Developer Network (SDN); customers and partners can engage with each other in our Enterprise Services Community to create new service definitions that meet their business requirements; Business Process Experts can engage with each other in our Business Process Expert community; and more. This community approach achieves tremendous results. For example, in our SDN community there are more than 5,500 postings per day and the average response time to a posted question is less than 18 minutes. Communities also provide SAP with the best possible way to reflect the voice of our customer in our own products. For example, in the latest SAP enhancement package, more than 50% of the enterprise service bundles delivered were developed via the SAP Ecosystem.

- The third implication is **the need to orchestrate co-innovation among the stakeholder participants.** The value of the ecosystem is maximized when its capabilities are linked. Through the ecosystem, we are bringing together diverse relationships, resources, and communities to help create the next generation of great solutions in concert with our own development efforts. For example, one of our Strategic SAP Technology Partners actively engages in the SAP Developer Network, the BPX community, and the High Technology Industry Value Network. When the company was about to release a new product, it worked with SAP and leveraged the ecosystem for product input, thereby leading to the development of a superior product with faster developer acceptance.

For SAP, the ecosystem is all about creating a valuable customer resource. This is an ongoing journey as we work with our customers and partners to create the best possible experience.

A Community-Authored Book

At its core, collaboration is a balance between emergence and control. If, as a sponsor of a community, you expect the community to only represent your views and priorities, you will be sadly disappointed. Communities have their own desires and will move—along their own timeline—in the direction they choose to go. If, on the other hand, as a sponsor of a community, you are afraid to "go first" and seed discussions, your community will not realize its collaborative potential.

This book represents a fascinating experiment that attempts to walk the fine line between orchestration and control by providing a seed for parts of a book and then seeking input from the community. Some parts of the book were first presented to the community in a finished state. Other parts of the book started out more skeletal. One chapter began as an empty shell. The community is still reacting to this content and commenting, changing, criticizing, and praising the content, all in a way that will ultimately improve the book.

As one of the leaders of efforts to promote collaboration in the broader SAP ecosystem, I have long wanted to see a thorough and complete explanation of the business process expert role. It appears that working together, BPX community members and SAP staff, along with professional researchers and writers, are creating a better book than could have ever been created by one of those groups working alone. We hope that you find the results of these efforts helpful and become moved to join in the process of making this content even better.

Zia Yusuf
Executive Vice President
Global Ecosystem & Partner Group, SAP AG

Preface

One thing we know for sure: All over the world people are calling themselves business process experts and are working to improve the quality and accelerate the pace of solutions created for their organizations. We know this because those people are using SAP's Business Process Expert community to explain what they do and to celebrate their successes.

What we were less sure of is exactly what business process experts are doing to achieve these results. However, during the creation of this book, more indicative trends have become clear. We know that business process experts are advocates for the business process perspective, the idea that a clear definition of a business process should be the foundation for creating efficient process and technology solutions. We know that business process experts help the business side communicate with the technology side. We know that they solve problems in requirements gathering, conflict resolution, training, and many other areas. But the reports coming in tell many tales. We hear of business process experts who are primarily communicators. We hear of others who represent the business point of view in the solution creation process. We hear of people who merely make up business process inventory to comply with regulatory standards. We hear of others who use modeling and visual development tools to prototype solutions or teach users to do this for themselves. These are merely facets of the business process expert role that have influenced the direction of this book.

Now that enough people have joined the party and a significant amount of experience has been gained, it has become possible to understand the many faces of the business process expert. This chapter attempts to look at the solution creation process as it is practiced in the modern world of Information Technology (IT) and identify the skills business process experts bring to the table and how those skills are combined into different types of roles.

To bring some order to this analysis we will proceed as follows:

- In Chapter 1, we will clear the decks with respect to the fundamental concept driving the business process expert, by examining the history of the term business process and the way that the concept has become central to the way business is conducted in the modern world. We will look at the business process reengineering movement, the way that enterprise software is driven by business processes, the drivers of business process standardization, and how all these currents have combined to lead to the definition of end-to-end business processes that span all companies participating in a business network.

- Next, we will examine how the business process expert is an advocate for the business process perspective, the notion that businesses should be driven and solution creation and automation should be guided by business process champions with a business process perspective. One central assumption of this book is that to be a business process expert in the most complete sense, you must be an advocate for the business process perspective and use that mode of thinking to guide everything you do.

- In Chapter 2, we will review the solution creation process to establish a clear view of the playing field that business process experts inhabit when, with a definition of a business process in hand, they turn to technology to help with automation and supporting functionality. We will review the problems that commonly occur in the solution creation process and the long-term trends that have led to the creation and evolution of the business process expert role.

- In Chapter 3, we will analyze the different personalities and styles of business process experts and the skills employed to improve the quality and speed of business process design and solution creation.
- In Chapter 4, we describe the SAP Business Process Expert community (BPX community). Started in 2006 and now over 400,000 members strong, the BPX community offers a rich panoply of resources for collaboration, which are explored in this chapter.
- By refining and helping to optimize business processes, business process experts become agents of change. Chapter 5 discusses the important area of organizational change management and the challenges faced by business process experts in helping organizations make often difficult changes for the better.
- Chapter 6 describes the technological toolkit of the business process expert, which includes Web 2.0 technologies such as wikis, blogs, twitter, and IM, as well as business process modeling tools.
- Chapter 7 describes patterns of success. This chapter identifies several types of patterns of success, including patterns that promote adoption of the business process expert role, patterns that promote creation of successful solution, and patterns for optimal communication, collaboration, and organizational structure. This chapter is largely a work in progress, though it already features a substantial contribution from the community.
- The book concludes with an Afterword that describes the process used to create this edition of the book and the plan for going forward. After all, the canonical version of this book is—even as you read this—on the BPX wiki. You can find it by visiting bpx.sap.com or simply by googling "Your Community BPX Book."

Contributions from the Community

Throughout the book, you'll find boxes that highlight content that was added via the wiki. Sometimes community members directly edited the text, and those changes became part of the body of the chapter. Sometimes members suggested the addition of substantive content; such suggestions are being held for the next edition of the book (which is already an ongoing effort on the wiki for this book).

This book was created in a hybrid fashion that combined research interviews and writing from the Editors and the Evolved Media team with comments on the content harvested from the community members about early versions of the content.

We would like to especially thank those members of the community who contributed, either directly or by interview:

John Alden
Axel Angeli
Yvonne Antonucci
Bob Austin
Kerry Brown
Richard Campione
Juan Carlos Carracedo
Paul Centen
Ajay Ganpat Chavan
Uwe Dittes
Mark Finnern
Anne Fish
Andre Fonseca
David Frankel
Tanya Furlan
Ginger Gatling
Mo Ghanem
Pinaki Ghosh
Swati Gokhale
Frauke Hassdenteufel
Sumarno Chandra Hie
Volker Hildebrand
Richard Hirsch
Phil Kisloff
Paul Kurchina
David Lincourt

Gail Lipschitz
Nancy Marguiles
Ashish Mehta
Alexander Obé
Kieran O'Connor
OwenPettiford
Marilyn Pratt
Jon Reed
Ann Rosenberg
Prasad Sammidi
Natascha Schuberth Thomson
Kelly Schwager
Jim Spath
Steve Strout
Matt Stultz
Helen Sunderland
Puneet Suppal
Paul Taylor
Arun Varadarajan
Jochen Vatter
Anbazhagan Sam Venkatesan
David Vonk
Aaron Williams
Mark Yolton
Zia Yusuf

Whether or not your name is listed here, we'd like to invite you to join in the conversation and comment on what you read here. Please visit the wiki for this book and help us improve it.

> ### Key Starting Points
> The BPX Community site (bpx.sap.com), which has many resources, among which is the wiki for this book. You can find it by clicking on wiki from bpx.sap.com or by googling "Your Community BPX Book."

1 The Business Process Perspective: Key to the Business Process Expert Role

Too often in discussions of the business process expert role, a precise definition of the term "business process" is never established. The term has been around for so many years and is used by so many people that everyone assumes a common understanding. In this book, we are going to start from the beginning and attempt to establish a clear definition of business process by looking at where the term came from and how it has bee — n used in different contexts. We realize that attempting to capture all connotations of a term like business process is beyond the scope of this book. Our purpose is simply to explain the meanings that are relevant to the business process expert role.

Reengineering and Business Processes

While the analysis of business processes dates back to Adam Smith, who in 1776 described the step-by-step division of labor in pin factories in *The Wealth of Nations*, the modern emphasis on the term dates from 1990 when Michael Hammer and James Champy introduced the concept of business process reengineering. Hammer and Champy define a process as "a collection of activities that takes one or more kinds of input and creates an output that is of value to the customer."

The essence of their approach was to encourage business executives to take a step back from the details of what was going on in their business and to think about what needed to be done to create value for their customer. The goal in performing this analysis is to increase efficiency and eliminate unnecessary work.

Reengineering proved extremely popular in the early 1990s in the US and other countries, as threats from global competition had increased the pressure to improve product quality and cut costs (remember the ISO 9000 craze?). The concept was adopted by some companies with noted success and criticized by others as attempting to enforce an overly mechanized view of how work gets done. By the mid 1990s, enthusiasm for the concept started to wane but the term business process remained an important part of the business lexicon.

> **From the Wiki: The Business Process Perspective**
>
> This chapter is new to this book (that is, it was not originally included on the wiki) but was born from comments received both on the wiki and in interviews that we needed a greater emphasis on how the business process expert champions the business process perspective. The authors would like to thank those who gave us this feedback, particularly David Frankel.

For the business process expert role, there are two observations that we think are relevant. The first is that the notion of taking a step back and focusing on the optimal design of the process for creating value **without regard to the technology** used to automate or support that process is a powerful first principle. This book uses the term "business process perspective" to refer to this way of thinking.

The second observation is that one of the barriers to the success of business process reengineering was in the flexibility of IT systems at the time. In the 1990s, the movement to implement ERP and other systems of record was just gaining steam. The configurability and flexibility of these systems was far more limited than the software in today's enterprise. In addition, support for collaboration and information management was a fraction of what it is today and far less emphasized. As a result, business process reengineering was held back significantly because technology itself could only play a small role in supporting the optimal process. The idea was sound; we believe it was just ahead of its time.

Business Processes and Enterprise Software

Enterprise software applications provide another important foundation for the business process expert role. Enterprise software solutions, like Enterprise Resource Planning (ERP), Customer Relationship Management (CRM), or Supply Chain Management (SCM), and others act as systems of record for the enterprise and record data that keeps track of the state of assets and activity. Applications also help automate business processes such as creating purchase orders, invoice processing, and order management and provide support for analyzing data and creating reports. Enterprise applications can also recognize and respond to various events that occur based on business activity, such as the expiration of a time limit for receipt of payment for an invoice, which may trigger a dunning letter.

In order for an enterprise application provided by a software vendor like SAP to succeed, it must be able to implement a process that is common to many companies and handle as much as possible of the differences between those processes by means of configuration. In other words, for an enterprise application to succeed, it must have embedded in it a deep understanding of a core business process and all the variations on that process. An accounting system, for example, must be able to handle US Generally Accepted Accounting Practices (GAAP) and European GAAP in the same system, switching on and off capabilities and applicable rules based on what country, division, or area the software solution is being used in.

In a sense, a business process expert plays the same role that the product managers and designers play in an enterprise software company. Business process experts must be able to look at the business processes in a company and assemble a solution that allows as much flexibility as possible to meet needs that are likely to occur. Flexibility isn't required everywhere. Rather, companies usually have a core set of value-creating processes that are the focus of innovation and optimization. The business process expert should play a leading role in helping business executives, enterprise architects, and the CIO craft the existing and future business process and solution landscape. The right combination of processes, solutions, and infrastructure will allow innovation and optimization to take place at the lowest possible cost and impact, just as well-designed enterprise applications allow customer needs to be met through configuration. In both cases, the optimal state occurs because the designers understood the scope of current and anticipated needs and prepared for them.

Success in the design of flexible and often cross-organizational business solutions is not the result of luck or accidents, but rather is the result of a deep understanding of the challenges faced by a business. Business process experts can learn from the general approach that enterprise software vendors like SAP take toward embedding business processes in their software. SAP for example has an integrated set of communities that gather high-level industry specific requirements (Industry Value Networks), design applicable code called web services (Enterprise Services Community), and then incorporate those services into the design of its products. SAP customers and partners participate in the Industry Value Networks and the Enterprise Services Community, bringing a wide breadth of experience to bear. SAP product managers also collect requirements from direct interaction with customers, through customer advisory councils, through user groups, and from analyst firms.

The process of gathering requirements and designing a product for a software vendor may involve hundreds or even thousands of people. It often takes place in cycles that span months or even years. Informally, business process experts can cast the same sort of wide net and imitate this process. They can systematically reach out to the people in the company who have knowledge and experience, like enterprise architects and executives, who are looking at defining the best path forward for the business.

In addition, business process experts must be well versed in the ways that enterprise software applications support business processes as they craft solutions that automate the optimal processes for a company. Of course, the optimal process may sometimes have to be compromised based on practical considerations. At times, it may make sense to take a step back from the optimal process and implement something that fits more easily with the existing capabilities of an enterprise software package, if the compromise does not greatly reduce the business value obtained.

The goal, of course, is to compromise as little as possible and implement solutions that are as close as possible to the current understanding of the optimal business process. Much of the excitement surrounding business process management (BPM), service-oriented architecture (SOA), and the associated tools for modeling business processes and building systems is driven by the hope that such techniques will reduce the gap between the optimal process and the capabilities of enterprise software to support it. The more knowledge business process experts have about the requirements for the optimal process, the capabilities of enterprise applications, and the potential for BPM and SOA, the better prepared they will be for playing a leading role in creating solutions that drive companies' future strategies and products.

Business Process Standardization

Another valuable source of inspiration for business process experts can be found in the way business processes have been standardized in various industries and functional areas. When you take your credit card around the world and find you can pay for dinner in Mumbai, buy a book on Amazon from your desk in Palo Alto, purchase tickets to the FC Bayern Munchen game in Munich, or rent a surf board in Costa Rica, all with a similar experience during the purchase no matter where you are, it is all based on implementation of standardized business processes and the business models that support them.

The world of standards is vast and complex and full of both positive and negative examples. The Internet is perhaps the biggest victory ever for technology standards, although the fact that your cell phone works in all the places you can use your credit card is not far behind. Supply chain business standards are, for instance, much more developed than other areas because the economics of global supply chains have brought a large return on investment for creating and adhering to these standards. In other industries, standardization has crept along slowly and not made much progress.

Standards are sometimes developed by government, other times by consortia of companies who are interested in improving efficiency, and other times by one powerful player, such as Wal-Mart, which has been pressing for adoption of a standardized approach to RFID by its suppliers.

Business process experts need to be able to interact with and understand standards for business processes at many different levels. From an industry perspective, it is obviously important to understand the applicable standards and employ them in solutions. Companies can also benefit from playing a role in setting or improving standards. Much of the activity in modern standards-setting is focused on the development of web service APIs that allow business to be transacted safely and securely across company boundaries.

Just as business process experts can imitate the designers of enterprise software, they can also apply the broader lessons of standards setting to their own companies and ecosystems of partners. One common pattern of standardization is applied to the financial rollup of acquired companies. Instead of converting acquired companies to a common ERP system, some conglomerates have defined a set of web services that answers queries for financial information required for statutory reporting. To become part of the financial rollup, the acquired company must implement these web services. Using this approach, acquisitions can be fully integrated into the financial reporting processes in a matter of weeks. Business process experts frequently find opportunities for similar standardization in their day-to-day work consulting at companies. As standardization increases, they find that their focus is far more about creating effective applications to achieve business results than achieving efficiencies through standards. In other words, standardization paves the way for innovation. For people who use only landlines, it is a lot more difficult to understand the benefits of an iPhone than it is for people who already have a cell phone.

The Business Process Perspective

It is all well and good to recommend that business process experts should be advocates for the business process perspective, the point of view that recommends first understanding what your ideal way of working should be, and then seeking support to make that happen using technology as well as cultural change. But this is just the beginning. A complete definition of the business process perspective adds more meat to this concept and addresses how a company must change, add, or remove inefficient processes, how communication must occur, and how solutions must evolve in order to reap the most benefits.

What we now offer is an initial attempt at fleshing out the concept of the business process perspective and making it more useful. Like any initial attempt at an ambitious concept, it is likely that this definition will spark disagreement. Our hope is that such disagreement leads to rapid incremental improvement of these ideas. That the canonical form of this book is on a wiki that is available to be changed and commented on should make it easier for everyone who has the passion to join this debate to enter the conversation and help push this idea in the right direction.

Our work so far with the SAP Business Process Expert community (BPX community for short) has led us to the following definition of the business process perspective:

- **The business process perspective means thinking in terms of the largest picture.** How does a business process being defined relate to other processes inside and outside the boundaries of the company? How does the work flow in and out of the process? What events and content must be communicated to and from other processes?
- **The business process perspective means thinking in terms of the process not the software implementation.** What are the inputs and outputs to the process and what transformation is taking place that creates value for the customer? What are the ideal steps involved, not the steps that have developed for historical reasons or because of convenience? What parts of the process can be implemented now and which may have to wait for later? How can we keep the ideal process in mind so we can move toward it as opportunities for improvement arise?

- **The business process perspective means documenting business processes so that everyone can see their role and how it fits into the big picture.** What is the grand plan for creating value? How is that strategy linked to the processes we perform everyday? How do we keep track of our progress and performance? What must each role performing a process know about what is happening before and after to do the best job? What must we know about our partners, and they about us, to do the best job for the customer? How can we document processes in ways that will be understandable by everyone involved?
- **The business process perspective means a commitment to continuous, incremental process improvement.** How can we test our vision of the ideal process against the experience we gain every day? How can we allow everyone involved to make suggestions for improvements? How can we institutionalize the process of taking incremental steps forward? How can we incorporate new technology capabilities that may support process improvements? How can we continuously identify processes that need further investment and processes that need to be removed? What evaluation framework should we use?
- **The business process perspective means a commitment to adapt the organization to support the process focus.** How should lines of authority and the flow of information be transformed to focus both management and staff on the efficient execution and optimization of the core processes of the business? What are the design principles behind the current organizational structure? How can change be managed in the least disruptive way possible? How will people be trained to play new roles?
- **The business process perspective means a commitment from senior management to work through challenges that are inherent in the transformation process.** How can management express its support for the vision of an organization centered around business processes? How will management be informed when the time is needed for intervention to maintain the course in the face of difficulties and obstacles? How will management help maintain momentum for change in the face of the inevitable frustrations

and shortfalls in performance involved in learning new ways of working?

Puneet Suppal, an SOA strategy architect at Capgemini, describes a company that is organized around the business process perspective as a business process enterprise. "If you are still looking at these things separately in different chunks, you're going to manage them in separate chunks," Suppal says. "There is going to be inefficient overlap of activity, or there are going to be gaps when one process may assume that something is being addressed in another. But if you look at the work entirely as a complete process, you would be able to reduce inefficiencies. You would be able to see how things move end to end. You'd identify where all your problems are, and you'd be able to manage the business better."

The benefit of adopting the business process perspective and becoming a business process enterprise is that each part of the end-to-end process understands its role and the significance of what it is doing. Everyone has visibility into what is going on up and down the chain and can make better decisions and take action to make processes work better. Such a situation will not occur by accident. It is the result of a detailed vision, crafted from a deep understanding of what a business needs to do to succeed and a recognition that the vision will be improved based on experience. Only by creating a wider consciousness in the enterprise of such a vision and taking a methodical approach to gradually moving toward that vision can a business reach its true potential. Further, as companies move increasingly beyond their own boundaries, the ability to incorporate those end-to-end processes throughout the value chain could initially be differentiating and create substantial value.

An Advocate for the Process Perspective

As an advocate for the business process perspective, it is not the primary task of business process experts to lead the process of transformation. Rather, it is their task to convince the organization that the business process perspective is the right way to organize a business so that many leaders can emerge from all levels of the organization and effect change to improve business results. In that respect, the first thing that a business process expert must do is, in essence, get him or herself elected as a credible representative of the business process perspective.

Let's imagine that a business process expert was running for the position of advocate for the business process perspective and had to convince other employees to elect him or her. Perhaps the best way to summarize the way that a business process expert would pursue such advocacy is to imagine what kind of speech he or she would give to an audience of voters to gain their support. Such a speech might sound like this:

> *My fellow employees, please allow me a moment to let you all know why I want to be elected as advocate of the business process perspective and what I plan to do if elected.*
>
> *First of all, I am not going to run a negative campaign. I know that some of you are cynical about change in general and technology in particular. I'm not going to dwell on past failures caused by applying technology without using the business process perspective.*
>
> *I am not going to blame IT staff who sometimes overpromised but underdelivered because they were excited about technology functionality but forgot to consider business value.*
>
> *I am not going to criticize people who think only in terms of existing applications and ignore what technologies like mashups and composite applications could do.*
>
> *I am not going to get mad at people who see everything in terms of development tools and turn every challenge into a custom development project in ABAP or Java or Ruby on Rails.*
>
> *I won't waste time worrying about perfectionists who invent ornate business processes. It is great to be ambitious, but we must always test requirements against experience.*
>
> *I especially am not going to enter into a tirade against people who resist change by protecting their turf or refusing to learn new techniques. We must have compassion for people who resist change because the world seems to make them suffer for their attitude.*

To dwell on such negativity would get in the way of the important work we have to do to move this company toward the business process perspective.

I have thrown my hat in the ring for the honor of helping advocate for the business process perspective because we are a good company on the way to being great. I know one thing for sure. We are not going to get there unless each and every person in this company knows what must be achieved and why. Everyone must know why their job is important and their place in the overall process.

It shouldn't take much faith to believe that adopting the business process perspective can help us. If there weren't commonality in business processes, enterprise software like ERP wouldn't have become billion dollar industries.

Looking at how technology standards like web services led to business process standards, the power of the business process perspective is clear. Amazon and eBay now take in almost as much revenue through their web service APIs as they do from their web sites.

Both enterprise software and standards are based on the business process perspective. Part of adopting the business process perspective means focusing on making our businesses run properly, not focusing on technology. After we understand what to do, we can focus on how to get technology to help. Too often, technology is so exciting that we let the "how" crowd out the "what."

Making the most of the business process perspective means understanding where we are with respect to business processes. In some areas, everyone is aware of each other, with core processes automated and the important data captured. We share information and optimize performance by tuning processes.

In other areas of our company, we have little awareness of the big picture. The processes were not consciously designed, but emerged by accident based on application capabilities or an inadequate understanding of the needs of the business. In these areas, we must take a step back and think about the right way to do business, looking at it from an outsider's perspective. We must communicate that vision, refine it, and incrementally move toward it.

The work of the business process expert is the art of the possible because the next step we take is rarely a leap to a full implementation of our vision. The only way to find out if our vision for the optimal process is correct is to try parts of it and see if our instincts were right. Much of the time, experience shows us that only part of our vision was needed.

I think this is the best time ever to be a business process expert. We have a rich technology foundation. Enterprise applications like ERP are in place and working well. Web services combine data and functionality into composite applications. Web 2.0 technologies like blogs and wikis create a rich fabric for collaboration. If we start with the process, we will all reap the benefits.

As excited as I am, we must be humble. If this were easy, everyone would have done it already. Sometimes we will be faced with seemingly impossible problems. There will be bumps in the road. Change is never easy and that's why learning how to change provides us with an advantage.

So, if you don't elect me, please elect somebody to champion the business process perspective. I think that changing our collective thinking to be focused on business process is the most important thing we can do to ensure our future success.

It is exciting to imagine that if we could just vote for someone to be an advocate for the business process perspective, our companies would just change for the better. Also the business process perspective acts as a solid basis. Processes do not change as often as technology does, especially in our agile dynamic global world. As we all know, it is much harder than that. The business process expert is just one actor on a stage filled with C-level executives, line of business managers, and business users on one side and IT managers, architects, and developers on the other. The business process experts, enterprise architects, business analysts, solution consultants, and project managers are in the middle, attempting to bridge the two worlds. Even if everyone on the stage is committed to the business process perspective, there is a whole other category of work to be done in the realm of creating solutions that support business processes that we will address in chapter 2.

The first mission of the business process expert, the one related to advocating for the business process perspective, tells us where we want to go. But getting there involves embedding this vision and the related automation processes in the real world, in real applications. This is where the business process expert plays a different role, one that is not an advocate for a new way of thinking but one that is based on the ability to use empathy and communicate across boundaries. Chapter 2 explores the role of the business process expert in the solution creation process.

2 The Solution Creation Process

The reason the business process expert role is needed in the solution creation process is two-fold. First, the tools for creating solutions have become vast and, because of that, often more complex and someone has to be expert in understanding the business potential of the functionality and how to apply the tools in a process-oriented fashion. Second, the business process expert must understand the requirements that the business is attempting to meet and communicate those to the technology side. In both of these roles, the business process expert uses empathy for technology and for business needs to make sure that communication flows as it should and solutions are built that actually help the business perform better.

Overview of the Solution Creation Process

To understand the second half of the business process expert role, we must take a closer look at how solutions are created and the problems that typically occur. Creating an accurate, generic, universally applicable model for solution creation is a tall order. What we hope to do with the following model is identify the most important activities and processes that take place during solution creation so that we can better focus our analysis of what business process experts do and how they do it. Our generic model of solution creation is shown in Figure 2-1 on page 17.

> **From the Wiki: The Solution Creation Process Summary**
>
> This solution creation process summary is truly a collaborative effort. Community members suggested the addition of business case building, user-centric design, and change management, among other topics, and added refinement to descriptions and challenges. Ginger Gatling suggested that the user feedback gathered at the end of the process should loop back as input into future solution design.

Here is a brief description of the activities and processes in the Solution Creation Model along with associated challenges.

Activity/Process	Description	Challenges
Communication	Communication must occur along many vectors between people who do not speak the same language and have the same motivations.	Making sure that everyone understands each other and takes the time to communicate and listen.
Failure Anatomist	Frequently a business process expert is called on because things have not gone well in previous attempts at creating solutions. The business process expert er can be more effective if the reasons for failure can be discovered.	Building trust to get the real story of what happened in the past. Addressing the external factors that may have led to failure.
Table is continued on page 18		

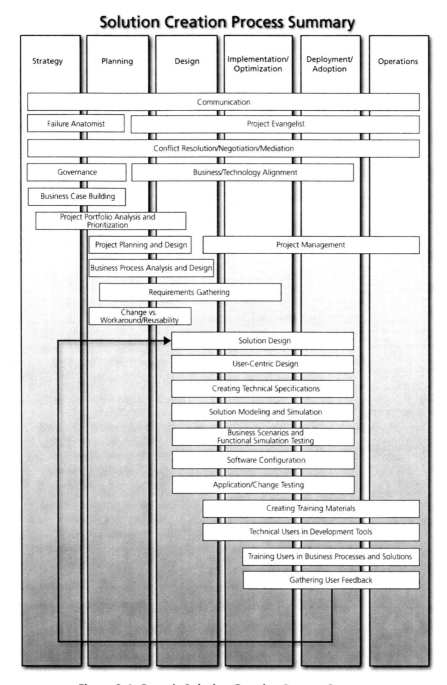

Figure 2-1: Generic Solution Creation Process Summary

Activity/Process	Description	Challenges
Project Evangelist	Business process experts are often the best positioned to communicate the value of a project to the rest of the company and prepare others for the roles they must play in supporting the project.	Generating excitement about the goals for the project. Motivating people to do what must be done to make the project successful.
Conflict Resolution, Negotiation, and Mediation	Tempers can flare when making decisions about what a solution will do and what it will not do. Disagreements can be based on deeply held values and interests.	Finding a way to come to agreement and move forward in a positive way, in which everyone feels fairly treated.
Governance	The solutions created must meet the needs of the business strategy of the company. They must also fit into an enterprise architecture development methodology.	Gaining the time of senior executives to make decisions about key issues. Ensuring that proposed projects fit with governance structure.
Business and Technology Alignment	The needs of the business for automation and information must be balanced against the difficulty of creating specific functionality.	Striking the right balance so that business requirements are met but large investments in functionality of dubious benefit are avoided.
Business Case Building	Business process experts may be called upon to help build a business case for a project.	Assessing the business case in light of how it would impact the company as a whole and what types of change it would bring

The table is continued on page 20

Even a quick look at the processes and activities involved in the solution creation process makes the importance of strong communication skills immediately clear. User requirements must be captured and communicated to technologists. Comments from users must be harvested. Knowledge must be transferred through training. Questions about the capabilities of software must be answered. In most of these cases, there is both a business and a technical aspect to the communication.

The other characteristic of these processes and activities is that they frequently require that business and technology considerations be understood and balanced. For example, it is usually better to do without a feature that is expensive to develop unless the need for that feature has been clearly established. Sometimes using standard approaches is not worth the sacrifices that must be made with respect to the optimal process. Other times new technologies cancel out gaps between the ideal processes and what was possible to implement. But to make these tradeoffs, you must understand both the potential business impact and the associated costs. Too often during solution creation, miscommunication results in a very expensive solution that does not meet business needs and is not technically sound for the long term.

Changing Nature of IT Solution Creation

Over the past 20 years, a series of trends have culminated in today's environment in which the business process expert role is thriving. To understand why the business process expert role is succeeding, we must understand the changes that have occurred in solution creation and the demands those changes are making on IT departments.

Development Methods: From the Waterfall to Agile

The traditional method of development is usually called the waterfall methodology because development proceeds forward in a cascading series of steps. In the typical waterfall process, systems are created or large projects are managed by starting with a requirements gathering process, then a system is designed based on those requirements, and then the system is implemented, tested, and deployed. This method has been used widely in all sorts of engineering domains.

Activity/Process	Description	Challenges
Project Portfolio Analysis and Prioritization	There are always more projects requested than capacity to fill them. The most important projects must be prioritized.	Understanding the business value, the investment, and risk associated with each project and providing qualitative and quantitative data needed for management to approve the projects..
Project Planning and Design	Projects must be constructed so that development proceeds with proper milestones, input is received from all interested parties, and the requirements for the solution are clear and agreed upon.	Making sure that technologists just don't take the requirements spec and run with it and then make decisions based on technical concerns and ignore how the solution will serve the business process.
Project Management	Making sure the project goes according to plan.	Keeping communication flowing and instilling an appropriate sense of urgency. Keeping project managers informed of technical interdependencies and adjustments needed to the project plan.
Business Process Analysis and Design	Defining, apart from the implementation, a business process that will serve the needs of the business.	Thinking in terms of end-to-end processes rather than the software that implements them. Engaging end users in the exploration of both in-system as well as off-system processes.
Requirements Gathering	Gathering from users the requirements that the solution must satisfy.	Separating the true business requirements as opposed to items that are nice to have.
The table continues on page 22		

> **From the Wiki: The Importance of Application Testing**
>
> Community member Ajay Ganpat Chavan highlights an activity in which the business process expert can play an important role: application testing. Depending on the impact of the change, the business process expert may get involved in preparing test plans, approving test plans, and even in the testing itself.

The waterfall method usually applied to creating IT solutions has been widely criticized and often leads to failure of the resulting solutions. It turns out that collecting requirements for a solution is far harder than it may seem. Much of the time the conversation about requirements between business and IT is dominated by technical considerations. Because the business side lacks the technical language to express requirements in sufficient detail, the conversation between business and technology about solutions is dominated by descriptions of functional capabilities leading to solutions that reflect the capabilities of the software at hand but do not solve the business problem. (A detailed description of this problem and a remedy that involves the application of systems thinking can be found in *Lost in Translation*, published by Evolved Technologist Press.)

Agile development methodologies are one of the responses to the failings of the waterfall methods. Agile development methods including eXtreme Programming, Scrum, the Unified Process, and others take the point of view that requirements are inherently unreliable. Instead of attempting to create requirements for a large system, agile methods seek to implement the smallest possible system that is useful and get it in the hands of users. Then a tight feedback loop between the users and developers allows the true requirements to emerge. The eXtreme Programming methodology, for example, suggests that those using the system and developers be sitting in the same room so that they can communicate directly. When this is not possible, sometimes a representative from the user community is situated with the developers so the feedback can be given rapidly.

In agile methods, development of a large system is done as a series of iterations in which functionality is gradually added. Each iteration adjusts the system based on user feedback, sort of like steering a car. In

Activity/Process	Description	Challenges
Change versus Workaround/ Reusability	Users, sponsors, and business community provide input about whether to create a solution or use a workaround. A change or workaround could be based on existing software such as ERP and Microsoft Office with some manual effort	Assessing costs and benefits and suggesting a way forward. Working on strategy to define a long-term approach.
Solution Design	Designing a system that supports a business process and the people playing roles in that process with information and automation.	Designing for flexibility, avoiding over automation, and serving the needs of the business process rather than letting technical considerations have too much impact.
User-Centric Design	Designing a solution from the user's perspective, not a system perspective	Discovering actual user needs versus making assumptions. Defining the right level of user to center on.
Creating Technical Specifications	The requirements for a solution must be specified in technical terms so that the solution can be implemented.	Creating a specification that is detailed enough so that the solution created will meet the business requirements.
Solution Modeling and Simulation	One method of designing solutions is to create models of the processes or to simulate the user interfaces in order to confirm that the design of the solution will meet the requirements.	Making sure modeling and simulation don't take up too much time and actually reflect the design of the solution. Making sure the feedback gained is used to improve the design.
Business Scenarios and Functional Simulation Testing	Documenting and describing business scenarios and functional simulations.	Making sure that the scenarios and simulations accurately represent the end state of the solution.

The table continues on page 24

some methods, like the Rational Unified Process, the interactions are focused on building the core parts of the system and confirming the requirements. Once that is done, a much larger effort takes place that is similar to the waterfall methods but that is based on requirements that have been vetted.

While agile methods are still controversial in some quarters, leading Internet companies like Google, Amazon, Yahoo, and others have shown that this method can work at scale. All of these companies have launched large software efforts in beta versions that are not yet finished and then used the experience gained in the market to improve the products. However, as is true for the business process expert role itself, if such a methodology is not incorporated companywide and sponsored and promoted at the corporate level, early indicators have shown that it is likely to fail.

User-Driven Innovation

Eric von Hippel, a professor at MIT, has become world famous because of his research findings that show the value of user-driven innovation. In two books, *Sources of Innovation* and *Democratizing Innovation*, von Hippel has analyzed how innovation takes place and demonstrated that when people who use products are given the means of production, they often create superior versions of those products that meet their needs far better than designs created by intermediaries. Von Hippel has shown that this effect holds true in a wide variety of industries and contexts.

Part of von Hippel's analysis points out that one of the reasons that user-driven innovation is so powerful is that it obviates the need to extract requirements from the minds of users into a form that can be understood by product designers. This translation process is seldom accurate. Von Hippel calls the user's knowledge of their requirements "sticky" because it is not easily extracted.

True user-driven innovation is limited to the scope of activities where the users of the products actually can develop new prototypes. In many situations, the means of design and production are too complicated for direct use by users. The success of user-driven innovation,

Activity/Process	Description	Challenges
Software Configuration	Most mature software systems have complicated configuration mechanisms that control their behavior and tailor functionality to meet different needs.	Software configuration is often a complicated undertaking performed by experts.
Application/ Change Testing	Preparing test plans and testing to ensure that a solution meets interaction goals and required objectives	Ensuring adequate time for testing given tight schedules. Allowing time for user testing.
Creating Training Materials	Training materials are needed to explain to users how the solution works and how they will interact with it to do their jobs.	Creating training materials that explain the context for the solution, not just step-by-step instructions.
Training Users in Development Tools	Increasingly, users are able to configure software to meet their own needs, much like they have done with spreadsheets for decades. Training is required in these new methods.	Providing access to the new development tools and monitoring the user-created solutions to make sure they meet standards for security and reliability and, if needed, scalability.
Training Users in Business Processes and Solutions	Once training materials have been created, users need to spend time learning the new solutions.	Motivating users to attend training. Making training sessions effective.
Gathering User Feedback	Once a solution is in place, user comments should be harvested as part of the process of improving the system.	Understanding what users really need, based on their comments about a solution, can be challenging.

however, has left many companies wondering how they can make this technique work for them. This remains a considerable challenge facing most organizations today.

Tacit Interactions

The first wave of enterprise software automated the processes that were common across most businesses. Then, processes that were common across companies in an industry were automated. For 30 years, with SAP and later Oracle leading the way, enterprise software started with financial, accounting, and control processes, which became known as Enterprise Resource Planning (ERP), then branched out to Customer Relationship Management (CRM), Supply Chain Management (SCM), Product Lifecycle Management (PLM), Human Capital Management (HCM), and Supplier Relationship Management (SRM). This collection of software grows every year with Governance, Risk, and Compliance (GRC) and Enterprise Performance Management (EPM) being the latest additions. Common functions, such as data warehousing and tools for integration, have been added. Standards for common business processes have been created. The result is today's world in which consumers and businesses buy millions of products every day without the knowledge that their transactions are eventually recorded and processed using SAP's ERP system.

What has been recognized in the past few years, however, is that the automation of ERP and the other sorts of enterprise software is only one part of the way that most people work. Most existing software supports transformational processes (those that change a physical good into something else) and transactional processes (those that follow a set of rules in a predictable or predetermined fashion). A new class of work called "tacit interactions" has been identified as key to productivity.[2] Tacit interactions are ad hoc tasks in which a knowledge worker pulls together information from many sources and then uses analytical tools or other methods to perform a task. Tacit interactions are complex combinations of judgment, problem-solving, and communication that

---

2   See "Competitive Advantage from Better Interactions" by Scott C. Beardsley, Bradford C. Johnson, and James M. Manyika, *McKinsey Quarterly*, 2006, Number 2 for more information on tacit interactions.

occur among spontaneously assembled groups and that happen differently each time.

Because of their ad hoc nature, tacit interactions are not predictable or automatable in the same way that enterprise processes are, but that does not mean that they cannot be supported or enabled by software. The trend toward service-oriented architecture has unlocked the information inside of enterprise applications, making it more readily available for use in tacit interactions and helping bridge the gap between end-user applications like spreadsheets and enterprise applications like ERP. The rise of Enterprise 2.0 collaboration (blogs, wikis, social networking, search, linking, and others) and mashups has also increased support for tacit interactions for those who understand how to use and configure those tools. Adding support for tacit interactions as solutions evolve will be a major challenge for business process experts for many years to come.

Enterprise 2.0 and Collaboration

The rise of user-created content and mass collaboration, using the power of the crowd, was dubbed Web 2.0 by O'Reilly Media, a leading technology publisher and event organizer. While investing momentum slowed down after the dot.com bust in 2000, the collective action of millions of people in the consumer space just kept on going, resulting in breakthroughs in many different areas. Blogs open the door for individuals to publish their opinions and glimpses of their lives to the entire world. Wikis are the basis of Wikipedia, the largest encyclopedia in the world, based on informal collaboration among a group of hundreds of thousands of contributors. Social networking sites like Facebook and MySpace allow people to record their social connections so that applications can report to others in their network on what people are doing. Applications also make use of those connections to share information and recommendations inside a personal network or do other useful things. Texting, instant messaging, and other wireless applications make all of these interactions mobile and real time. YouTube opened the door to mass distribution of user-created video. In addition, the development of services such as search, email, calendaring, document creation, document sharing, and networked storage are all increasingly available and reliable.

The consumer space authoritatively took the ball of innovation in computing. People in corporations were left wondering why their computing environments at work were so much more limited than what they could do as consumers on the Internet. What happened next is a phenomenon known as shadow IT. Just as personal computers and web sites infiltrated the enterprise by users working outside of the control of IT departments, blogs, wikis, and other parts of the collaborative dial tone became often adopted in the same way. People used technologies on the Internet, in the shadows as it were, to meet their needs. It didn't take long for IT departments to catch on and to start formally introducing Web 2.0 technologies into the enterprise. Harvard Business School professor Andrew McAfee dubbed corporate use of Web 2.0-style collaboration Enterprise 2.0.

The result of the rise of Enterprise 2.0 is that a new set of capabilities are introduced to a population of end-users who vary widely in their ability to accept them and make use of them. These capabilities in many cases are proving ideal for supporting tacit interactions and ad hoc collaboration of the sort that previously only occurred through long trails of emails. At many companies, younger workers seem to adopt and make use of Enterprise 2.0 capabilities much faster than older workers.

Widgets, Mashups, BPM, and SOA

Part of the revolution of Web 2.0 and Enterprise 2.0 is a phenomenon that O'Reilly Media called "innovation in assembly." This innovation is fueled by web services that provide access to public services like Google Maps as well as data and functionality from ERP and other internal enterprise systems as well. The browser has become a much more powerful user interface and application development platform due to the emergence of technology platforms such as Ajax, which allow browser-based interfaces to access data from many web services and process that data.

Environments are emerging that allow users to assemble new combinations of information and applications for themselves. Yahoo!, Google, Microsoft, Apple, Nokia, and others have widget frameworks in which chunks of information and functionality can be assembled on pages by users as needed. Simplified environments are emerging for creating mashup applications, another name for composite applications assembled from a variety of services. More complicated mashups can be assembled through the use of Business Process Management environments that provide mechanisms for visually describing business processes.

All of this innovation in assembly opens the door for users to create interfaces that have the information they want and applications that meet their needs. But, again, this is only possible if users can master the tools used to create such applications. Younger and more technically savvy users have been leading the way.

Staffing the Gaps: The Business Process Expert Role in Creating Solutions

So, looking back on the nature of solution creation and the trends shaping the modern computing environment, we can make several observations:

- All of the challenges related to solution creation that have been with us, and will be with us forever, must be addressed on a day-to-day basis. Activities such as governance, project portfolio analysis, training, communicating effectively, and organizational change management must be performed to a high standard to ensure success.
- Even when guided by the business process perspective, communication between the business side about their needs and technologists who create solutions has been discovered to be incredibly difficult. Agile methods make the assumption that such communication is almost impossible and experimentation through iterative development must be used to determine accurate requirements. Likewise, the research into user-driven innovation shows that information about user requirements is sticky and hard to translate into a form technologists can use to build solutions.

- User-driven innovation is on the rise as more and more capabilities are provided for people to craft their own solutions through means such as Enterprise 2.0 and mashups. But, at the present time, many of these mechanisms are too difficult for all but a small number of people to use. Furthermore, this fast increase in smaller shadow solutions could possibly create a counterproductive effect for the end-to-end business process.
- Support for tacit interactions is improving and Enterprise 2.0-style collaboration is on the uptake by certain groups of users in the enterprise. Web services are allowing corporate information to be brought into the mix, but the nature of the required dial tone needed to promote collaboration in the enterprise is still forming.
- The nature of the IT department is changing. In addition to building large systems that automate the operations of the enterprise, IT departments are now in the business of observing what users are doing for themselves and promoting proper standards when an ad hoc solution becomes popular. IT departments also seem to be moving gradually to a greater business understanding as the business organization becomes more at ease with IT savvy.

So, adding all of this together, it is clear that IT has left the shore of past practices and is on a journey to a new land in which users will do more for themselves, in which a mature and robust collaborative dial tone for Enterprise 2.0 will be available to people who know how to use it to support tacit interactions, and in which the difficulty of communication will be acknowledged and appropriate efforts will be made to overcome that difficulty.

But, it is not at all clear how long this journey will take. And no business can afford to wait around and use past practices in the meantime. Urgent questions must be answered during the journey:
- How can the business process perspective guide the solution creation process and make it more effective?
- How can maximum progress be made while the world of IT is learning to use these new tools and to adapt to new practices?
- How can companies do a better job of addressing the traditional challenges of solution creation?

- How can learning and solution adoption be accelerated?
- Who is responsible for debugging the process of solution creation at your organization?

There is an answer to these questions. It is a person with special skills, most prominently a sophisticated form of empathy and an understanding of the business process perspective, further described in Chapter 3.

3 Completing the Definition of the Business Process Expert Role

Now that that we have described some of the major challenges of solution creation, we can add to our definition of the business process expert.

The role of the business process expert in supporting solution creation can most simply and accurately be defined as a person with the ability to quickly understand business needs and translate that understanding into a form that leads to the creation of better solutions.

- Sometimes that understanding is translated into better requirements documents
- Sometimes it is translated into user-created solutions
- Sometimes it becomes suggestions for the way to use collaboration to help a project
- Sometimes it becomes innovation in new processes
- Sometimes it becomes conflict resolution
- Sometimes it becomes education and training

But in all cases, the business process expert is someone who has a special skill, that of being able to deeply understand the needs of the business and then to crystallize that understanding in a way that empowers others, reduces risk, and increases the quality of solutions. In many cases, we believe that the business process expert is the hybrid evolution of what used to be called a business analyst, a solution consultant, and an application consultant. Many business process experts will report to someone called a business process architect, who is in turn yet another level of business process expert.

> **From the Wiki: A Definition of Business Process Expert**
> The business process expert can most simply and accurately be defined as a person with the ability to quickly understand business needs and translate that understanding into a form that leads to the creation/composition of better solutions.

What the business process expert does when supporting solution creation is to staff the gaps that exist between the current capabilities of people and technology that stand in the way of achieving the vision of flexible, responsive IT that many observers see on the horizon.

Depending on the size, needs, process maturity, use of standards, and technological architecture of the organization and its industry, the business process expert's responsibilities will vary to one degree or the other. For instance, it may be that the business process expert works to optimize existing processes or, instead, to innovate entirely new processes. If the business process expert is at a large industrial company, the focus may be more on improving existing processes. At growing companies that are still working out their business models, business process experts may push the envelope of possibilities by creating entirely new processes. At most firms, the role of business process expert as innovator will grow organically as the benefits of accelerated business process innovation are recognized. Regardless of the size of the organization, to be effective, a business process expert must be a polymath and possess a host of talents. For example, most business process experts have a deep understanding of a broad range of existing business processes, software applications, UIs, and architectural paradigms such as SOA. Business process experts may excel at working with tools for collaboration, business process modeling, or visual application development. Most especially, the business process expert needs proven experience in a variety of practical business settings.

Types of Business Process Experts and the Skills They Employ

In the past, duties of the business process expert with respect to solution creation were performed in part by multiple roles, including business analysts, business architects, business consultants, implementation consultants, IT managers, and enterprise architects, to name just a few. Today, the well-rounded business process expert can be expected to possess the ability and talent of any number of the aforementioned roles and thus avoid some of the pitfalls they encountered working in isolation from one another.

> **From the Wiki: The Business Process Expert Role as a Hybrid**
>
> According to Bob Austin of Atos Origin UK, the BPX role is a hybrid of:
> - Business analyst, concerned with gathering business requirements
> - Business process modeler, concerned with developing process models using tool such as ARIS, SAP Solution Manager, and SAP NetWeaver Visual Composer
> - Application consultant, concerned with mapping business processes with applications
>
> The business process expert will also typically develop business process-based metrics, analytics, and dashboards.

The sophisticated empathy that we suggest is the foundation of the business process expert's ability to help create better solutions may be applied in many different ways. Our research has shown that each successful business process expert we have encountered has a certain talent for empathy and combines that talent with other skills as needed. Each business process expert finds a way to make a difference based on his or her unique skills.

While there are no hard and fast rules, our research has identified five general categories of business process expert beyond the advocate for the business process perspective that illustrate different ways that empathy creates value:

- Business process expert as Organizational Change Manager and Therapist
- Business process expert as Requirements and Process Analyst
- Business process expert as Solution Designer
- Business process expert as Empowerment Coach
- Business process expert as Innovator

Each of these five types of business process experts uses the ability for empathy and rapid understanding to apply different skills in different areas of the solution creation process.

Business Process Expert as Organizational Therapist

Organizational and communication problems are at the root of many failures to properly create IT solutions. When the business process expert acts as an organizational therapist, she frequently arrives when progress has reached a standstill and both the business and technology staff involved realize they need help getting unstuck.

It is no coincidence that, time and again, both business and IT professionals, along with business process experts themselves, describe human dynamics as being the single most difficult obstacle that organizations encounter when striving to improve.

Helen Sunderland, a pioneer business process expert who has been involved in many of the early discussions of the nature of the business process expert role, frequently talks of the way she "helicopters in" to help solve problems. The metaphor is an apt one because it includes the idea of arriving with a view from above to solve an urgent problem, perhaps a crisis.

Without exception, every business process expert and related professional we talked to while putting this book together testified that the greatest challenge to success in creating new solutions and optimizing business processes is change management. Why? Richard Hirsch, a business process expert whose areas of expertise include knowledge management, portal integration, and others says, "People have the tendency to resist changing their current workflow unless the benefit of change is made obvious." (We will delve more deeply into this issue in Chapter 5 on organizational change management and the business process expert.) Regardless of the situation or the players, business process experts need deft hands, keen eyes, and sensitive ears if they intend to win the client's confidence. A mastery of the following soft skills is key to delivering the help that organizations need to become more effective.

Skill: Communication

Helen Sunderland has noted in her blogs on the SAP Business Process Expert community (BPX community for short) that nearly 60% of her time with clients is spent engaged in varying levels of communication efforts. "Clear and honest communication," she says, "is key between a business process expert and any client, but particularly essential in a short-term assignment which results in ultimately influencing business change.... Your ability to communicate clearly in a manner appropriate to your listeners is essential to your success.... You may consider yourself a guru in your area of specialty, but no amount of sage wisdom will suffice to make up for your lack of ability to communicate well with multiple audiences." Claiming too much expertise could actually turn people off and make them feel that the business process expert is not listening but instead trying to take control.

Whether business process experts are working in house or have been hired as outside consultants, from an organization's perspective, they are expected to assume the role of a "trusted advisor," Sunderland says. Business process experts are sophisticated professionals whose success at any given assignment depends not only on their business and IT skills, but, just as importantly, on the means by which they convey their credibility and authority to the people with whom they work. No matter how brilliant and creative the business process expert, without the ability to garner a client's trust, he will not be able to overcome the client's reluctance to move forward with the new ideas and information the business process expert is there to provide. Usually one or more members of an organization will be familiar with the business process expert's background. Be that as it may, however, such knowledge will never preclude the necessity to establish and nurture personal, face-to-face relationships as a means of increasing the productivity of every interaction.

Here is a tried and true method for establishing trust and credibility: after briefly outlining their history and concentrating on past experiences in positions as a "trusted advisor," business process experts should always exhibit openness and candor, clarifying for the client that there are no secret agendas and that they are the number one priority.

Since communication consists of the *exchange* of information between individuals, the business process expert's ability to listen is just as important as the ability to speak and write well. After they have established some initial credibility with their audience, it is incumbent upon business process experts to ask a number of specifically directed questions and then to listen with great attentiveness while the client responds. The client may be brand new, or the business process expert may have already established a good working relationship. In either case, with each new assignment, it is important that business process experts behave as if they know nothing about the client and their needs. Remember, the business process expert's objective is to sort out the critical issues from the "noise" that surrounds them. Remaining simultaneously objective and fully engaged is the quickest way to see through to the core issues, formulate the big picture (a holistic view that includes the organization's culture, political landscape, business, and IT practices), and ultimately, devise recommendations for optimal solutions.

An alert ear goes a long way in this regard, as does finely honed intuition. While clients may well be explicit in their difficulties, needs, concerns, anxieties, and fears, it is often the case that many of their most crucial issues will remain cloaked, revealed, if at all, between the lines or through various hints. When this is the case, the business process expert can ask further questions for clarity and tease out further information. While on the surface it may appear that a client's circumstances have nothing to do with the business process expert's own experience, frequently they are similar at heart. According to Sunderland, a bona fide technique for business process experts to quickly understand their client's environment and issues is to listen for similarities to their own experiences.

The audience with whom business process experts must communicate includes a host of players, from engineers and programmers in IT, to HR and quality control managers, and, finally—and most importantly—to various stakeholders and members across the C suite, without whose ardent sponsorship meaningful growth would be impossible. When communication between business and IT is successful, a creative energy flow develops that produces value for both the organization and its customers.

Political, cultural, and business matters are more likely to arise organically at first than are technical matters. For business process experts to avoid unwanted surprises, such as a discovery near the ostensible completion of a project that a key standalone application was "hiding" beyond the main system's landscape, they must ask every question they can to ensure that that they have identified all of the nonconforming systems, along with the principle members of the cast.

Once business process experts have grasped the big picture, troubleshot the problems, and formulated suggestions for improvement, they must deliver their conclusions in crisp, concise language that is catered to the client at hand. For instance, Sunderland suggests that if finance is the audience, business process experts will best succeed by giving them a no-nonsense message that articulates the desired cost benefits. If, on the other hand, the message is directed toward members of the C suite, it ought to be delivered in unambiguous business language. If the audience is a technical or configuration team, it is essential that the message express all solutions and configurations using technological language and jargon.

Skill: Facilitation and Mediation

Traditionally, a cultural and ideological gulf has separated professionals of the business and technological sectors. In many organizations, an environment exists where businesspeople look upon technologists as mere tools for building greater capital while technologists scorn the businesspeople for their indifference to anything that doesn't affect the bottom line.

As this book will repeatedly emphasize, the business process expert's ability to coexist in both the worlds of business and technology is a strength that organizations would do well to employ by way of transforming what is a potentially toxic dynamic into a culture of collaboration that benefits both sides of the equation and, in the end, the business itself.

When business process experts enter any new situation, among the many roles they may assume is that of a mediator between two or more parties holding potentially different viewpoints and concerns with the intention of facilitating improved relations between them. To gain a clear picture of a company's existing culture, business process experts can start by asking a number of questions. What, for example, is business's current relationship with IT? What is their level of process maturity? Is there any animosity, rivalry, or competition? If so, what are the various possible solutions for alleviating the tension or purging it entirely? From a business point of view, what would be the ideal role of the IT department? What would business like IT to do for them today? On the flip side, how does IT perceive business, and how can business best serve IT?

The empathy that we have described as being paramount to the business process expert's success is crucial in such a context. The goal is to remain objective enough to avoid taking sides while simultaneously understanding and identifying with the pain points of all concerned parties. Business process experts must be able to tell each individual or group that they hear their frustration, anger, anxiety, and fear, and then explain why and how these problems cannot do anything but obstruct growth for the company and for the individuals that run it. It is critical that patterns of blame be identified and rooted out. It is just as critical, too, to avoid concentrating on IT as the source of every problem.

Once they have located the sources of the company's ills, business process experts must then communicate ways to transform existing patterns and processes in ways that are both healthy and acceptable to all concerned, working with any unavoidable constraints, such as lack of funds, time, and space.

Conversely, as the old saying goes, business process experts should never try to fix what is not broken. It is vital that business process experts recognize and continue to use those aspects of a company that are perfectly functional. Clearly, the organization does many things well. Otherwise, it would not be in business. The key is to refrain from changing for the sake of change and to isolate those components that are inefficient or broken from those that are working well.

Skill: Change Management

Change management involves the process of actually transforming an organization once the direction has been set. Business process experts frequently fill gaps that exist in the design and execution of transformation efforts. This can involve anything from aiding in business case development,to helping create a strategy for performing the transformation, to explaining the roles that everyone will play in a transformation, to helping with troubleshooting during project execution. Frequently during this stage, business process experts recruit people who are involved in the project that they have identified as being effective agents of change.

> **From the Wiki: A Change Management Dialog**
>
> The BPX wiki includes a project that involves a pilot for a fictional company, Big Machines Corp. This project helps business process experts understand the types of interactions they may encounter, as illustrated through scripts. The page on change management, by Owen Pettiford, Richard Hirsch, and Hans Butenschøn, is especially relevant to this discussion. See https://wiki.sdn.sap.com:443/wiki/display/BPX/The+role+of+Change+Management.

In applying all of these soft skills, the business process expert does what many therapists see as their core mission: Bringing problems and negative patterns to light so that an enlightened awareness of the big picture is created. In this way, the healthy parts of the organization are expanded and brought to bear to help heal unhealthy parts. Business process experts generally find that the role of organizational therapist can pay large dividends as more and more people are brought to a deeper understanding of the goals that an organization is attempting to pursue. This approach especially bears fruit in organizations that have divisional silos.

Business Process Expert as Requirements and Business Process Analyst

Programs of change and transformation need specific targets. Business process experts play a crucial role in solving problems inherent in gathering requirements and designing processes, as well as in a necessary precursor to that step: evaluating strategy and building the business case related to those requirements. In a perfect world in which user-driven innovation is easier, the challenges associated with requirements gathering would be less onerous. Indeed, gathering requirements for many solutions would be unnecessary because users could create what they needed for themselves. Processes would be modeled in such a way that they could be easily adjusted and optimized on the fly. But we are not there yet and, even in such a world, some solutions will be too large or too broad to be created by users without the help of technology professionals. Challenges of requirements gathering and process design will always be with us in some form. Right now, in most organizations, gathering requirements and designing effective business processes to meet them remain intractable problems that business process experts are stepping in to solve.

Requirements Gathering

Requirements are essentially lists of things that a system should and should not do. The challenge in most requirements processes is that it is hard to find the right level of detail. Requirements that are too broad could be satisfied in many ways, some of which may not achieve the business goals. Requirements that are too specific mean that the technologists' role of finding the best way to meet the requirements with available technology is limited. Ideally, requirements explain the goals that users are seeking to achieve, as well as various boundaries on an acceptable solution.

In most cases, requirements are gathered in some sort of document that explains the business goals for a solution, which features or functions are absolutely needed, and which would be nice to have. In the best case, requirements documents extract the "sticky" knowledge, as described in Chapter 2, from users about how they do their jobs and what they need to do them better.

Unfortunately, for reasons that are presented in detail in the book *Lost in Translation* (Evolved Technologist Press, 2007), analysis of requirements for solutions frequently is dominated by IT considerations. Because technologists are so comfortable with abstractions and complexity and are also, in general, enthusiastic about the development tools and software applications they use, discussions of detailed business requirements are frequently dominated by information about the relative merits of functionality. The "how" crowds out the "what." *Lost in Translation* provides a concrete way to implement the business process perspective by focusing on five aspects of an information system—the values, policies, events, content, and trust relationships (VPEC-T). Through these dimensions, the business users can express the personality of the system they want created to help the business.

While few business process experts have codified the way their empathy works to extract "sticky" knowledge and translate it into a comprehensive document, almost every business process expert we talked to reported that representing users in the requirements process was one of the key ways they added value. It is important to remember that any requirements document will fall short of answering all the questions that technologists have when building a solution. One of the reasons that the VPEC-T approach works is that the principles of the information system at the foundation of a solution are described so that technologists have detailed guidance about designing a solution that will work. Business process experts tend to be involved throughout solution creation, answering questions and providing guidance all through the process so that the needs of the users are well represented and the "what" of the solution is not overcome by the "how" of the technology.

> **From the Wiki: The Value of Prototypes**
>
> You can shorten the requirements gathering phase by making good use of prototypes. Furthermore, prototypes allow the user to validate what is being created at an early stage. In essence, the business process expert can ask the user, "Is this what you need?" and the user can confirm the way forward, speeding development cycles and preventing requirements from being lost in translation in many cases.

BPM and Top-down Process Design and Modeling

Almost all IT solutions are created to serve the needs of a business process. Sometimes those business processes are precisely defined in a step-by-step manner. Other times they are looser, more collaborative processes that have end and beginning states but not a lot of detail about how the work in between will be performed. Business process experts report that one of the key ways they accelerate the creation of solutions and improve their quality is through exercising their skills and good taste in designing business processes that often cross organizational boundaries. Frequently, business process experts add the most value when a company is starting over and designing or redesigning the way they do things from the top down, as recommended by the business process perspective. Business process experts may take this opportunity to start to introduce some of the relevant findings from Business Process Management (BPM) research and best practices.

David Frankel is an expert in Model Driven Architecture. When asked what he perceived to be the business process expert's critical skill set, he said, "A solid understanding of the company's business processes, good judgment in distinguishing automatable processes versus tacit interactions, an ability to identify existing IT components available to support business processes, and an ability to define requirements for new IT components where they are needed. The [business process expert] must also have the ability and mindset to model business processes precisely and should be familiar with the ideas behind model-driven software development. The [business process expert] must have the ability to communicate effectively with and understand the point of view of business people and IT people."

To design and automate a business process requires an understanding of its definition, the business goals that an organization wants to achieve, and how available technology can provide the solution. Marilyn Pratt, a leading business process expert evangelist and blogger for the SAP BPX community, defines a business process as "a set of activities transforming a defined business input into a defined business result." A business process step, on the other hand, is "a task or an interaction performed by a process component either with or without human interaction and together with other steps form a business process." As for a business application, it is "a collection of business processes required to address specific business needs, implemented via a set of application and software components running on a platform." A good example of this is SAP Customer Relationship Management (SAP CRM). Finally, a business process platform is "the combination of SAP's Application Platform with SAP's technology platform; it supports the creation, enhancement, and seamless execution of business processes and business scenarios."

Yvonne Antonucci, Associate Professor in the Department of MIS and Decision Sciences at Widener University, suggests, in her BPX community blog, that adopting a BPM approach can yield a variety of benefits when designing processes. First, Antonucci says, after organizations recognize that "BPM must focus on a holistic view of process management . . . [they] should focus on the goal of integrating management, organizational issues, people, process, compliance, and technology for both operational and strategic activities with resulting business processes that produce value, serve customers, and generate income. These strategic activities encompass analytical and predictive methods with technologies in an effort to create agile organizations. While well-defined and automated processes can be successful accomplishments for organizations, sustained success lies in the ability to create value through effectively managing and orchestrating these processes across the organization."

Second, noting that a comprehensive skill set is clearly required to create a successful BPM program, Antonucci echoes our former suggestion that "an examination and definition of the BPM practice is required."

According to her investigations, the general lifecycle of an organization's BPM program begins with process planning and strategy that people use to direct the analysis, design, and modeling of business processes. In turn, these models form the basis of the configuration, implementation, and execution of the processes themselves. Once the processes are in place, they must be monitored and controlled. The ensuing data is then analyzed again, and the results used to refine the processes still further. These steps provide the basis for a feedback loop in which the analysis, design, modeling, configuration, implementation, and execution of processes are continuously improved upon.

While many companies are in fact establishing basic BPM programs, what they generally lack, Antonucci says, are "the management practices of key success factors that must also be integrated in order to encompass both the business and IT aspects of BPM." These success factors include strong executive as well as divisional process sponsorship, clearly identified process owners, employee incentives for process improvement, and proven change management, among others. As we noted earlier, until business process experts arrived on the scene, these success factors accrued only slowly, over time, and were focused on either IT or on business, but rarely on both. And, since they were not typically integrated, they were usually inconsistent. To guarantee success, business process experts must be deployed with the skills that integrate process and management practices for processes that run across an organization's entire ecosystem.

This assessment of the current state of BPM is the foundation upon which the business process expert's comprehensive skill set can best be defined and improved. For now, it entails a variety of hard and soft skills, including a holistic understanding of business and technology; strong SOA, business process, and application skills; and knowledge of both formal modeling and do-it-yourself (DIY) tools.

In other words, the design of business processes is no longer just a matter of setting down the goals, the beginning and end state, and a list of tasks to be performed. Business Process Management has become a discipline that can be used to better organize the entire scope of activities of a company. Business process experts act as a force that brings order, one solution at a time, by sorting activities into different categories and defining the boundaries between automation and human activity.

> **From the Wiki: Business Process Modeling with BPMN**
>
> Why is business process modeling important and what do you need to know? If you googled that question, you would likely find references to courses by Dr. Bruce Silver, a leading independent BPM industry analyst. Dr. Silver has provided the BPX community with some exclusive content: a 6-part series on business process modeling with BPMN. This vital series is available both in eLearning (https://wiki.sdn.sap.com:443/wiki/x/2_E) and article (https://wiki.sdn.sap.com:443/wiki/x/NwAB) formats.

Tacit Interactions, Bottom up Refactoring, and Optimization

In addition to the top-down design of business processes, business process experts play a bottom-up role that amounts to battlefield management. When in the process of helping smaller projects and troubleshooting, business process experts frequently find ways to improve the efficiency or execution of business processes through small adjustments.

One way that business process experts achieve results from incremental changes is when they recognize a situation in which a tacit interaction can be more effectively supported.

David Frankel points out that certain aspects of business processes are transactional, repetitive, and fully automatable. Other aspects are not. In many situations, knowledge workers use unique tools to negotiate, collaborate, and then exercise on the ground judgment after synthesizing information. In a tacit interaction, knowledge workers pull information from a number of sources and try to think about it and make a decision.

Tacit interactions include coordination, monitoring, negotiating, identifying new markets and value propositions, and the like. The percentage of employee activities that constitute tacit interactions is increasing across the board in all industries and has reached as high as 60 to 70 percent in the financial and healthcare sectors.[2]

Here are some examples of tacit interactions:

- Pulling together information to decide what companies to partner with on a strategic initiative
- Coordinating product development, field support, and partners
- Negotiating an OEM deal

Such tasks cannot be automated because they aren't sufficiently regimented.

In the world of Enterprise 2.0, business process experts frequently are able to suggest ways of using collaborative tools such as blogs, wikis, and mashups—especially those that use web services to access information and functionality in enterprise applications—to more effectively support tacit interactions. In such cases, the tacit interaction moves from completely informal mechanisms, such as email and spreadsheets, to one with just a bit more structure in a shareable space, which now allows others to join in the process and creates a reusable repository of knowledge.

Business process experts also frequently suggest incremental changes in processes that allow unused parts of enterprise applications to be brought to bear on processes that were taking place with insufficient information. Incremental changes to the design of processes can also remove roadblocks and increase efficiency. But to perform such changes confidently and effectively requires a broad perspective, something that business process experts consciously cultivate.

---

[2] See "Competitive Advantage from Better Interactions" by Scott C. Beardsley, Bradford C. Johnson, and James M. Manyika, *McKinsey Quarterly*, 2006, Number 2 for more information on tacit interactions

Business Process Expert as Solution Designer

While we have so far emphasized the soft and fuzzy, empathetic, business process nature of the business process expert, it is important to realize that they are down and dirty technologists as well. Like all technologists, they tend to love details and complexity. A big part of the impact of the business process expert role has come from the way these folks bring this detailed knowledge to the table during solution creation.

The business process experts we have spoken to report that they apply detailed knowledge about the following areas as they do their jobs:
- User-focused process design and modeling
- Application capabilities and boundaries
- Application configuration
- SOA, mashups, and integration

User-Focused Process Design and Modeling

Earlier, we described how business process experts play the role of requirements analysts and process designers. In this role, business process experts look at both the big picture from the top-down and at how processes can be improved incrementally from the bottom up. There is another level that business process experts operate on that overlaps with the process design skills already described. Business process experts are quick studies and have a large amount of knowledge from the user perspective. In the design and analysis role, business process experts act as advisors and representatives of those involved in managing and overseeing the process, but in the role we are about to describe, business process experts act as counselors and advocates for the users who are executing the processes. In this regard, the business process experts are superusers who not only see the big picture but who are also conversant with all of the knobs and dials that are used to carry out the work. The main overlap with the process design role is the gray area between suggesting an incremental improvement in a process and suggesting a change to user interface that may fall short of being considered a change to a process.

The main work done by business process experts is to again apply empathy, but this time by sitting in the chair of the workers who are making the process happen. All too often, solutions are imposed on workers who are not involved in the process of describing the solution. One of the reasons that agile development methods work is that they increase the information flow from users during the solution creation process. In addition to helping to advocate for users, business process experts often perform the role of facilitating user acceptance testing during the rollout process.

Much of the time, business process experts playing the role of user advocates are suggesting ways of improving user interfaces or adding new screens that offer relevant contextual information to improve decision making. But as more development is model-driven, business process experts are often the ones who convert an existing application to a model-driven form. Solutions based on this sort of modeling usually pave the way for user-driven innovation down the road. But the first step, that of creating the model-driven application from scratch, is something that usually must be done by a superuser.

David Frankel sees a convergence in the future between the modeling used at high levels to describe business processes and that done at the level of application creation for model-driven development. "Those of us who are pushing model-driven techniques in software development are pushing from the bottom up to that business/IT intersection," said Frankel. "On the other hand, you've got business process management going on. BPM is bringing automated tooling to the managing of business processes. You are increasingly able to model processes at a high level in a machine-readable way and have the processes execute. BPM is moving top down toward that intersection of business and IT. And it became clear to me that here we are, both from the bottom up and from the top down, pushing toward this intersection and that we are going to collide and work at cross purposes unless these communities start having a dialogue." Frankel sees business process experts as one of the important participants in this dialogue and attempts to bridge the gaps between these two worlds in his blogs and speaking engagements.

Application Capabilities and Boundaries

One of the key roles business process experts play is that of application experts, people who have a deep knowledge of both the application's capabilities and its place in the overall application portfolio. In this capacity, they know the functionality of specific applications and how that automation is properly applied. The key to success in the use of vendor-created software is knowing what the software was intended to do—forcing software to do something it was not built for can be a recipe for disaster. On the other hand, today's software frequently has untapped capabilities used only by application experts. The business process expert can step in to provide help in exploiting application capabilities to their full extent.

Business process experts contextualize this expertise because they are, by definition, experts in the business processes that a company intends to implement. When the business process expert combines the knowledge of what applications can do with the understanding of what processes a business needs to implement, she can then start separating which parts of the process will be automated by a vendor-provided application, which will be supported by collaboration tools in a tacit interaction, and which will take place in the brains of the workers. She can determine which parts of the process will take place in several applications that may be working together. In this way, the business process expert draws boundaries and carves the process into chunks and then allocates them appropriately.

Helen Sunderland finds that her interaction starts by finding out the set of tools that are currently in place to perform a business process. "Clients tell me they have been doing something with Excel spreadsheets. They then ask me to help them design something that may be a combination of Excel spreadsheets and SAP products, or all SAP products, or all Excel, or whatever it is—we plan on cocktail napkins. They want me to talk to them about what we're doing but keep wearing my SAP hat at the same time so that I can provide them with the best solution."

In performing this analysis, Sunderland keeps her eye on the big picture. "We don't ever work with business process in isolation. Yes, most of the time technology helps make the process more effective, but sometimes there are also people issues, logistical issues, and strategy issues that need to be addressed. One issue can't be discussed in isolation. Everything needs to be discussed and understood together."

Sunderland said that drawing the boundaries and allocating parts of the process to the right application involves combining an understanding of best practices with the ways that work for the company. When this knowledge is combined with an understanding of what a vendor product can do, the chances of success improve dramatically.

"What I do is I build planning applications so that managers can enter their expenses, so the CFO can forecast his financial statements, so that the sales planners can figure out their sales projections. I build budgeting software forecasting tools, the whole gamut," said Sunderland. "And because nothing's out of the box—everybody has some standard processes that they already use in planning. Everybody plans their salaries; everybody plans their operating expenses in some fashion according to their chart of accounts. However, they may not have tools that are effective for the audience that is using them. So you can put a very sophisticated planning tool on a manager's desktop, and perhaps he's out on the shop floor in some factory, in some no-name town, and he's not really that computer-literate. You've failed because you haven't presented to the end user a final product that meets their needs. You want a very sophisticated application on your financial analyst's desktop, but you want something very simple and easy to use on a casual user's desktop who hardly ever logs on."

Sunderland is also careful about being too aggressive about forcing users into best practice processes. "I actually have a real issue with the phrase 'best practice' because I don't think there is such a thing. I think there are 'leading practices,' but I don't think there's a 'best practice.' What are the best questions you can ask so that you can then apply your experience to those things? The business process expert role is one that's primarily about designing business processes. You have to stay on top of the technology in order to make sure it's effectively implemented."

From Application Configuration to Solution Composition

Once a clear understanding of which application will do which part of a process has been established, the challenge is to make each application do what it is supposed to. Modern software was built to be configurable. What this means in practice is that each software product comes with thousands of settings and configuration mechanisms that control the behavior of all the various aspects of the software.

> **From the Wiki: SOA Skills**
>
> As described in the next section, on top of configuration, we have the flexibility offered by service-oriented architecture, which can in effect decompose applications into small sets of consumable, reusable services. Composing, configuring, and reusing what is available has become a critical skill in itself.

Ginger Gatling, an active member of the BPX community, sees configuration of applications as a separate body of knowledge that must be mastered to make the most out of a vendor-provided application. "Configuration is complicated enough that you really need to have deep knowledge available. And even though it's not technical knowledge, it's not necessarily business process knowledge either. It's sort of application knowledge. It's not application knowledge about what you want the business process to do. It's more like quasi-technical application knowledge about configuration. When we create a purchase requisition, is it going to become a purchase order immediately or does it need approval? If it needs approval, is it going to go through workflow or are they going to go look for it? Hundreds of decisions like these need to be made."

Gatling sees the business process expert as the legacy that should be left behind after the expert team from systems integrators that may have been involved in configuring the software when it was originally installed has departed.

"But if you look at this business process expert as partnering with the business organization to really understand the processes, how they're supposed to work in the company, and the technology that SAP has to extend those processes, the business process expert role will stay in-house and could be the legacy left by system integrators.

SOA, Mashups, and Integration

The rise in availability of web services and mashups that combine services into new applications has greatly expanded the possibilities for tailoring functionality to meet the needs of specific knowledge workers. A mashup, another name for a composite application, can combine services from many different sources to provide exactly the information and functionality needed. Services can also be the foundation for integrating applications so that information from one application can be synchronized with another application. In this way, for example, the customer information gathered in a CRM application can flow into an ERP application.

Business process experts play a key role in leading the brainstorming, design, and implementation of mashups and service-based integrations. An increasing number of easy-to-use tools can help to create mashups and configure service-based integrations. But, in order to effectively use these tools, you have to understand how business processes are being implemented and the role that each knowledge worker plays in these business processes. Because business process experts have such knowledge, they are well equipped to design mashups. Bear in mind that mashups not only help move processes along but also provide a pulse on processes through, for example, dashboards. Additionally, business process experts can guide customers through the ever-more populous landscape of new mashups that are being developed every day.

Business Process Expert as Empowerment Coach

Business process experts find fellow travelers everywhere they go—people who have a desire to help build better solutions but who need help in focusing their efforts. One of the most commonly reported roles in the research for this book so far is that of coach. Business process experts teach others what they know about how to help build solutions. While the scope of such activities is not limited to one area, the business process experts included in our research reported that most of the knowledge transfer and empowerment took place in the following areas.

- **Requirements Gathering**: Business process experts often find people who share their talent for empathy and have strong communication skills. Once these people learn the ropes, they can easily become business process experts themselves, especially with regard to capturing requirements and communicating those requirements to technologists. So far, business process experts have reported that it seems easier to train someone with business knowledge in technical skills to be a business process expert than the other way around.
- **Modeling**: The process of describing solutions at all levels using models is gaining popularity. Frequently, the people who best understand the business processes find modeling tools strange and difficult. Business process experts report that they have been able to accelerate progress in creating a comprehensive set of process models by training motivated business people in modeling tools. Sometimes these modeling tools are used at high levels to model end-to-end processes that span the enterprise. Other times, these tools are used to model workflows using mechanisms like SAP Guided Procedures, which are commonly used to guide smaller segments of business processes but which can be used to model end-to-end processes as well.
- **Collaboration**: Business process experts often know just when to put a blog or wiki into place to make communication happen. Often, people are timid about introducing such tools because they don't know what will happen once information is out in the open. If an organization has a collaborative dial tone, that is, an infrastructure supports collaboration, business process experts can lead the way and show people how to put that dial tone to good use.
- **Configuration**: Configuring software is all too often considered a black art. But for decades software vendors have been working on making configuration as easy as possible, and in some cases they have succeeded. Business process experts report that some business users are ready to dive in and learn how to configure software

to meet their needs. When this happens, the door to user-driven innovation is unlocked.
- **Development Tools**: Frameworks for assembling user-interface widgets, environments for creating mashups, and model-driven visual programming tools have all removed much of the complexity of developing certain types of applications. These tools, while much easier to use than the previous generation of development environments, still have personalities and quirks of their own and take some getting used to. Business process experts report that with a little training to overcome the awkwardness of these tools, some users catch fire and become leaders in creating new solutions.

Business Process Expert as Innovator

One of the privileges of being involved in so many different parts of the organization and in the solution creation process is the opportunity to gain a wide perspective. Business process experts tend to dive in and become intimate with both the work of individuals and the workings of technology. This breadth and depth often leads to ideas for new processes and new solutions that fit into the big picture in unexpected ways. The final role we will examine is that of the business process expert as innovator.

The business process expert will usually play the role of innovator in two ways: as primary innovator and as a catalyst for innovation.

In the role of **primary innovator**, the business process expert will have an explicit responsibility for innovation as part of the job description. In some companies, this fits into a formal innovation framework in which certain parts of the company are tasked with finding innovations that optimize current business processes, that are adjacent to existing lines of business, or that are outside of the bounds of the current operations of the company but which have the potential to be thoroughly disruptive, following Harvard professor Clayton M. Christensen's meaning of that term. The role of primary innovator is difficult to carry out. Innovation in most cases is something that emerges unexpectedly and is not predictable. To be tasked finding something new and valuable on a regular basis requires contact with many different parts of an organization. Some companies put the responsibility for innovation on every employee, and in such cases a business process expert who is a primary innovator may be responsible for harvesting all of the ideas that bubble up and running them through an evaluation process.

In the role of **catalyst for innovation**, a business process expert is responsible for recognizing opportunities for innovation and bringing them to the attention of the appropriate people. In this mode, innovation is not a primary responsibility but something to be on the lookout for. If a company has a formal process for innovation, the ideas that are recognized should be submitted through it. Otherwise, the responsible parties in the line of business or IT department should be notified.

The five types of business process expert just described cover most of the activities that were described in the research so far. While most of the business process experts we talked to had skills in one or two of the categories described, nobody acted as a business process expert in all five categories. This suggests that, to be a successful business process expert, you must focus on a specific role that is well-suited to the skills you have as well as to the needs of the organization.

Summary: The Value of Adopting the Business Process Expert Role

The value of adopting the business process expert role primarily accrues from accelerating the creation of IT solutions that are of high quality and that serve the needs of the core processes of business success. When the business process expert is a powerful advocate for the business process perspective, and then speeds progress toward support of the right processes through empathy, dramatic change can take place. Bear in mind that this is not purely an evolution of the business analyst role since the skills required are much more than those of a business analyst and an application/implementation consultant combined. However, there is reason to believe that those roles are well-positioned to evolve into business process experts.

> **From the Wiki: Show Me the Value**
>
> BPX community member Jon Reed raises a question that is on everyone's mind: how, in times of downward rate pressure, to quantify and justify the value of the business process expert. Clients may push back and ask consultants not to "get fancy," says Reed.
>
> Furthermore, can the business process expert role be outsourced? Since business process experts facilitate communication and solution creation, the answer is quite probably no. Other skills can be commoditized, but business process experts bring a value-added factor in helping foster solutions that enable efficient and effective business processes. Effective communication cannot be outsourced.
>
> If you've found a good way to quantify the value derived from BPX-driven projects, please contribute what you've learned on the wiki for this book, which is linked from bpx.sdn.com. Specifically, we are interested in learning how can a business process expert convince the CFO that adopting a business process approach will save a certain amount of the company's IT investment in the next 10 years, a savings that would be lost if the solution were implemented without a business process approach.

Business process experts seem to act as a lubricant for communication and a catalyst for skill development and empowerment. In essence, business process experts staff the gaps that exist in most organizations between the business and technology sides. By focusing on the known problems, project meltdowns can be avoided. Here are a few ways that companies that have adopted the role have expressed the value of the business process expert:

- The rate of failure in creating and deploying solutions to support business processes drops
- Solutions are created faster
- Companies become faster and more agile
- Communication and cooperation between IT and the business side are improved
- More value is extracted from technology currently in place

- Technology is aligned much more precisely to serve business needs

The rest of this book covers topics that will speed your journey toward making good use of the business process expert role in your organization. Please visit the wiki and report back what you have learned so that we can further improve this book.

4 The Business Process Expert Community

The business process expert role and the SAP Business Process Expert (BPX) community (bpx.sap.com) are inseparable. The community was started as a way to explore the role and to provide a place for sharing leading practices around business process and solution design, but it also helped define, propagate, and evolve the business process expert role. The vibrant participation of hundreds of thousands of people and the exponential growth in membership indicates that the business process expert role is indeed a welcome description of what businesses need and want to institute to make a difference for their customers. Every month, more business process professionals use the community to be more effective, find answers, and share their knowledge. In this chapter, we will look at:

- How the community got started
- The technology capabilities and organizational basics of the community
- How people use the community to get their jobs done
- Where the community is going

By understanding the history and scope of what business process experts are doing inside the community, people who are not yet involved will see the benefits of participating in the community, while community members may find new ways to utilize the community for their personal development.

Axel Angeli, community member and founder of Logos! Informatik, stated that community is essential: "No successful movement in the world can work without a strong community backing it up. If you want to evangelize the ideas of business process experts, you have to have people who speak the same language, who support your ideas and spread the word. You cannot have one person to spread the gospel. All organizations that go out there and try to bring good ideas in the world work in communities. If I go into a company and say, "I'm the expert," and a question comes up and I cannot ask somebody else whom I trust, then I'm not a good expert. So community is an essential component of every successful management strategy. Evangelism doesn't work without community."

How the Business Process Expert Community Got Started

The SAP Business Process Expert community (BPX community for short) was founded in response to an intuitive, felt need. It seemed that there was a role that people were playing that bridges the domains of business and technology. To see if this was in fact the case, back in 2005, the founders of the BPX community conducted surveys in the SAP Developer Network (SDN), inquiring into the nature of the roles people played in their work. They learned that roughly 20 percent of the SDN community found it difficult to provide exact descriptions of their jobs that related to the technical world; their work resided deeper in the business world. About half of their work went toward meeting business objectives while the other half was directed at achieving IT goals. This discovery inspired the formation of the SAP Business Process Expert community as a grassroots initiative.

The BPX community was established to create an environment to help business process professionals alike to share, learn, and adapt to the rapidly changing technology and business landscape. The main idea was to use the power of the crowd to accelerate learning, connecting, and sharing business process-related information for the benefit of customers.

The goals of the community were ambitious. The founders wanted to help define and support the role of the business process expert, as well as to promote the business process perspective and the adoption of the business process platform as a paradigm for enterprise computing. Following the example of SDN, and using the same infrastructure, the BPX community created a venue for sharing and capturing knowledge through blogs, wikis, eLearning, webinars, and a content library. (In fact, partly as a result of the growth of the BPX community, the original SDN platform has changed its name from the SAP Developer Network to the SAP Community Network, which currently encompasses three communities: SDN, BPX, and Business Objects.)

The BPX community should be considered a collection of communities with many specific topic communities under it. It is like a country with many cities where people form subgroups to accomplish their goals and share information and interests.

> **From the Wiki: The Value of Community**
>
> Richard Hirsch says, "The BPX community provides me with a means to meet other people who have similar interests. The community helps me a lot because it gives me the ability to ask questions and talk to others who are dealing with the problems I am. We all have an arena where we can test ideas and behavior patterns."

Today, the BPX community provides a strong social online Web 2.0 network for business process professionals such as business analysts and application and solution consultants. Members range from members of the public to customers to partners to SAP employees who are evolving into business process experts. The community enhances members' work lives by providing a platform to connect with peers and share business process ideas, find answers for related business process and system implementation challenges, and find the best approach to new or existing processes. The community now has many industry communities where business process experts can share their knowledge about the specific industry challenges and trends they face.

To illustrate its rapid growth, the BPX community in 2007 had just 3 industry forums; it now has 18 and continues to expand organically. In addition, the community has groups where people can find answers and share knowledge about functional topics such as Enterprise Resource Planning, Customer Relationship Management, Supplier Relationship Management, Supply Chain Management, Governance Risk and Compliance, organizational change management, and so forth. In 2008, the BPX community became the designated place where members would find all documentation of the Enhancement Packages to SAP ERP, the regular stream of updates to the SAP Business Suite that includes the Enterprise Services bundles, sets of web services that provide access to data and functionality that can be used to create widgets, mashups, and composite applications to support business processes.

Embracing Web 2.0 technologies, the community leverages the creativity and expertise of its diverse membership to deliver many benefits to all. Customers can delve into product offerings provided by SAP and its partners, resolve implementation challenges, identify talent, and learn about industry and process best practices. Business process professionals are mastering effective implementation strategies, gaining community-wide visibility, and growing their businesses by promoting their expertise. Internal SAP experts and product mangers are communicating with customers to provide information about their solutions and gather business requirements to improve SAP product offerings. Through all of this collaboration, the BPX community supports business process and solution success.

The Technology and Organizational Structure of the Business Process Expert Community

The BPX community is based on the same technology infrastructure as the SAP Developer Network. The community works, like most online communities, because technology supports the interaction among thousands of people who connect with each other and continue to interact both through the community and through other means. A major amplifier of community activity is the regular meetings between community members that take place at SAP events like SAPPHIRE and TechEd, as well as at other local gatherings inspired by mentors and top contributors. At these face-to-face meetings, people complete the introductions that began online and create lasting and fruitful business relationships as well as friendships.

> **From the Wiki: Online and Face to Face**
>
> Matt Stultz, Home Depot's director of SAP technology, states that the business process experts who are emerging within Home Depot make regular use of the industry communities. Home Depot hosts local chapter events in Georgia, Stultz says, that are regularly attended by over 300 professionals. According to Stultz, the events have been a great mechanism for involving people who might not otherwise be aware of SAP organizations. The benefit of this setting is plain: people with varying skill sets and knowledge gather to share what they know in the hopes that new, unforeseen ideas might result.

Here are some of the main collaboration capabilities of the BPX community that support interaction among its members:

- **Community Home Pages**: The BPX community has many community home pages and section pages that highlight content and that are refreshed like a weekly newspaper. These pages relate to specific industries, topics, and areas of interest such as enterprise SOA and business process modeling. They provide a good first stop to find out what's going on related to a specific interest.

- **Forums**: The discussion forums of the BPX community allow people to ask questions and get answers from other community members, to bring up ideas for discussion, and to generally broadcast ideas and collect responses. Forum questions range from general questions that are asked over and over again ("How do I get started as a business process expert?") to the specific and highly functional ("How do I set up General Ledger according to Chinese compliance laws"). Many members find that forums are frequently used as a starting point for their involvement in the community and become their favorite resource.

- **Wikis**: The BPX community has made fascinating use of wikis for many different purposes. The canonical version of this book, for example, is located on one of the BPX wikis. Other projects on the wiki include a series of scripts created by members to explore the situations that face business process experts in the workplace, as well as training materials about how to become a business process expert advocate and how to use modeling tools. You'll also find analysis of key performance indicators (KPIs) and documentation of end-to-end scenarios. All of the content on the wikis is available for editing and commenting.

> **From the Wiki: How Human Interaction Affects Business Steps and Processes**
>
> Richard Hirsch speaks very highly of a venture on the BPX community wiki that provides a setting for community members to engage in a virtual project. Once everyone has logged on, each person adopts virtual roles, including the business process expert, the business analyst, and the enterprise architect. After that, they all work together to solve a common problem via instant messaging. The resulting scripts are posted on the wiki.

- **Blogs**: People with something to say—whether SAP staff, partners, companies that use SAP, consultants, analysts, or others—all put their thoughts out for consideration on blogs that appear on the community. Lively discussions are frequently sparked on the comments section of provocative blogs, and it is not uncommon for one blog to spark many others.
- **Articles**: The articles section is a shared content library that includes contributions from all the communities on the SAP Community Network (SCN). The repository mostly contains PDF articles, white papers, and presentations that have been uploaded from various sources. Articles can be found by keyword search and by navigating the tags assigned to the articles, as well as via the home pages and the article archive. Blogs and wikis are frequently created as guides to collections of articles.
- **eLearning and Podcasts**: eLearning and podcasts that cover technology and best practices of all kinds are available on the BPX

community. These modules provide audio commentary of slide presentations as well as audio and video sessions. The eLearning library is shared with all the communities on SCN and can be navigated by tags.

- **Collaborative Workspaces (cw.sap.com)**: These are private areas for collaboration limited to a smaller group working on confidential material, like new innovative business processes, or white space analysis, or BPX roundtables. The nature of this collaboration is private and it normally operates only for a limited time, until information is ready for prime time in the public areas of the BPX community. Furthermore, this part of the community is not yet hosted on the SAP Community Network platform but is in its own separate area. By the end of 2008, this will be integrated within the SCN platform.

These are merely some of the capabilities that comprise the current repository of business process expert content and the means by which people find each other and communicate. Further acceleration of activity on the community takes place through the evangelism of community staff, mentors, and evangelists. People like myself, Marilyn Pratt, Richard Hirsch, Helen Sunderland, Jon Reed, Puneet Suppal, Paul Taylor, Anne Fish and many others (forgive us if we did not mention you by name) play the role of editors, evangelists, writers, and moderators who suggest ideas, make connections, sponsor projects, organize events, answer questions, and generally do whatever it takes to make the community work for other members.

How the Business Process Expert Community Is Used

Online communities often seem like plates of spaghetti. If you pick up one strand, you never know which others may be connected. You may start with a forum question, get some information, see an insightful answer, look up the blog written by that person, and find that she has written an article that introduces you to new ideas. Section pages, blogs, wikis, forums, eLearning are all entry points that can lead in multiple directions.

While the connections between the content in the community may be sprawling and unpredictable, many different patterns have emerged for how people use the community to solve their problems and connect with peers.

Finding Answers, Sharing Knowledge

Perhaps the most common activity in the BPX community, and pretty much every other online community, takes place today in forums and in the comment sections of blogs and on wiki pages. Chatting, tweeting, and instant messaging are still used in a monolithic way but trends indicate that these technologies are slowly integrating with online community platforms. Someone puts out a question in search of knowledge and others step up to the plate and share their knowledge. This can happen amazingly fast. The most popular queries are answered within about 20 minutes. The benefit to the person asking the question is clear: getting an answer and learning from the experience of others. The benefit to the person providing the answer is also clear. The motive is often philanthropical, and it may have the side effect of enhancing the responders' reputation. You can also learn from answering questions and be stimulated to think new thoughts.

Alexander Obé, moderator of the BPX Governance, Risk, and Compliance community and SAP Customer Advisor, sees sharing knowledge as an online extension of natural human behavior, "The web technology allows interaction worldwide. Ten years ago, you had to attend a conference. Twenty years ago, we had news groups and that was similar to what we have now with forums. People have an innate willingness to share and support each other in similar quests."

Collaboration about questions and answers takes place across boundaries that are sometimes not crossed in other domains. It is not uncommon for people from competing companies to help each other find needed information. Sometimes this happens because people don't know who they are talking to. Other times it happens even though people know they are working with a competitor, because the community is a safe place to focus on craft and best practices.

In a way, the creation of online communities like the BPX community has seen a shift in the way we find—and sometimes how we pay for—knowledge. It used to be a value among technologists to keep secrets since their knowledge was a key to job security. Today, the value is sharing, and much knowledge that used to be hard won through long experience is now available quickly through the forums, blogs, wikis, articles, and eLearning. You can use this ability to find answers and sometimes to evaluate knowledge workers such as consultants. If they tell you something, you can then check it against what the community tells you and improve your ability to evaluate how someone will be able to help.

"Sharing on the BPX web site allows spreading information much faster and much more effectively," said Alexander Obé. "It somewhat devalues the guru knowledge, the special knowledge, the special experience, because there's a range of very experienced people who are willing to share. So, where 10 years ago you paid $2,000 for somebody who has a special expertise, that special expertise is now sometimes freely available to anybody who understands the basics of how to use the BPX community." It should be noted that we live in an age of increasing information. While more information is freely available today, the application of specialized knowledge to particular business problems is still well worth paying for, in many cases.

The Intellectual Property Dilemma

Often we get questions about why people would share their intellectual property: the business processes and solutions that form the basis of how they make money. Several points should be considered. In many cases, it could be that the knowledge is generally known so that in its essence, it is not really intellectual property. First, consider the knowledge about standard business processes that made the creation of standard software suites possible by the likes of SAP, Oracle, PeopleSoft, Siebel, and others. Second, it may be that the person believes that sharing a portion of their knowledge or expertise will create an appetite for more and thus that sharing becomes, in essence, a product sample that leads to more business. It just takes a few experts to start the fire and join a dialog about a topic, a dialog with many possibilities. Third, as the world becomes increasingly complex, and more mashups and solutions are created (even beyond the boundaries of your own company), knowledge of the business process is no longer the key differentiator for many professionals. Instead, they must know how to take advantage of business process knowledge and select or design the appropriate solutions for meeting customers' needs.

In a discussion with Richard Campione, who heads the business suite solution management organization at SAP, another important topic about sharing knowledge was raised. Why would SAP and its ecosystem open up a public community that shares important and related solution knowledge with, for instance, competitors? Richard made a quick reference to the Cold War. The former USSR kept its knowledge contained and private and in the end lost the Cold War, whereas the US opened up its gates of knowledge to be shared collectively and, in so doing, made faster advances than the USSR and came out the winner. It is believed that the SAP ecosystem will grow faster by openly sharing most knowledge than it would if it would it kept this knowledge a secret.

Phil Kisloff, an SAP mentor, explains it similarly. "The economic benefit, to a company or a business, of giving freely of your time or sharing insights or experiences is, in a way, the same economics of the open source movement. At a point long ago passed, not being part of the project became more costly than being on the inside, in terms of net opportunity for learning lost."

Finding People to Help

The BPX community helps people find and connect with others who have expertise to solve business problems. Many of the people answering questions on the BPX community are consultants for whom knowledge sharing acts as a form of personal visibility. People often find others inside their own organizations, as well as beyond the borders of their companies, who are working on similar projects and who are discussing their challenges or answering questions on the BPX community. The BPX community is full of people who would be called "connectors" according to the taxonomy of social networking found in Malcolm Gladwell's book, *The Tipping Point*. Frequently, a request for assistance of some kind is met with a suggestion by a connector to contact another person who has the sought-after expertise.

Training

Once an individual, a company, or a consulting firm has committed to the business process expert role, or even found parts of the approach or the associated domain knowledge useful, the community can be a powerful and often free resource for training. The content in the blogs, wikis, forums, and articles is organized and freshly updated by many experienced and knowledgeable community members. If you are looking for an introduction to business process modeling, you will find guides to all sorts of aspects of the topic, from high-level introductions to swimlane diagram usage to deep dives into business and technology issues. Most of the time, these guides are part narrative and part an assembly of references to other content in the community. This is all fueled by a powerful search engine that continues to improve.

At the level of the individual, the community is a goldmine for people who are seeking to improve their skills and keep their knowledge fresh. As Jon Reed, author of the *SAP Consultant Handbook* who moderates "Ask the Expert" on SAP careers for searchSAP, says, "The seismic plates of SAP skill sets are shifting, but the opportunity is there, like never before, to actively seek to fill any gaps in your own skill set."

Many new members of the BPX community are functional and application consultants who want to evolve their SAP application-specific skills in a specialty area. The community is able to help with this by guiding conversations through a collection of forums such as the one that is currently available for issues related to organizational change management. When forums are created for specific purposes in the BPX community, they ensure that the issues raised there are pertinent to those who join the forum. These forums are constantly being pruned toward this end, says Marilyn Pratt, community evangelist for the BPX community. For example, one organizational change management forum, with two dedicated moderators, has been created from a conglomerate of other similar but frequently unrelated forums, and then trimmed down to guarantee that the threads posted to it address change management from a purely technical standpoint.

Co-Innovation

SAP is promoting co-innovation with partners in its ecosystem and with its customers through various initiatives. One example is the Industry Value Networks that bring together select groups of SAP, industry-leading customers, knowledgeable partners, and other experts to examine the forces driving an industry to better determine the technology needs to solve white space business problems. The private Enterprise Services Community, part of the BPX community, helps design collections of enterprise services that are implemented as part of SAP products or submitted to standards organizations. SAP's Co-innovation centers provide an environment for SAP customers, partners, and startups to develop ideas and implement prototypes. On the BPX community, there are numerous projects to create content that have involved self-selected groups of community members. All of the features of the BPX community support co-innovation, but the Collaboration Workspace, the private space for BPX roundtable collaboration, has proven especially successful of late.

The BPX community supports co-innovation by acting as a sounding board that can bring people to the table to help develop and improve ideas. When David Lincourt, Vice President, Field Services in SAP's Global Defense Industry Business Unit, developed a new widget, the BPX community provided a means for him to spread the word about it in ways that he had not foreseen. By announcing it in the community, professionals whose presence he was unaware of responded with interest. Soon, others were engaged, more network links were created, and momentum grew. In the end, the value of the widget's functionality sold itself. Without the evangelistic assistance of others, along with the buzz that was created through it, Lincourt believes the project would have taken much longer to get off the ground. When new users saw the widget through the eyes of those who were already inspired by it, they, too, reacted with excitement.

> **From the Wiki: Co-Innovation with Business KPIs**
>
> A project that has direct relevance for helping assess the value of the business process expert is the "KPI project," as it is affectionately known in the BPX community. The Business KPIs wiki is the place for ongoing quantification of key performance indicators, which can help provide metrics by which to judge the success of initiatives. Visit the Business KPIs wiki on the BPX community at: https://wiki.sdn.sap.com:443/wiki/x/x4Y for links to videos, Second Life, and a wealth of information about KPIs.

Knowledge Capture

In the equivalent of the cry from Andy Rooney and Judy Garland musicals, "Let's put on a show," business process experts have similarly declared, "Let's create knowledge." One project involved business process experts playing different roles in IM sessions and then capturing and analyzing the interaction. Individual contributors have started the ball rolling with a blog post or a wiki page and then others have joined in. Of course, as we all know, not every project really takes off, but, the ability to start a project with nothing more than a good idea means that much more genuine collaboration and knowledge sharing can occur.

Identifying and Confirming Trends

The hive of activity in the community often reveals new directions in thinking and activity before they have become mainstream. Alexander Obé finds that his position as a moderator of a section of the BPX community provides him with a sense of what's happening next. Axel Angeli feels that a community is necessary to understand trends, stating that "anticipating trends can only happen in a community because you sense something is going on and then you confirm it with others." Regular visitors to the BPX community are rewarded with a preview of issues that are coming into wider circulation.

Understanding and Defining the Business Process Expert Role

Since its inception, the BPX community has been intrigued with understanding how and why the business process expert role works.

The project to create this book is just one example, but the forums and blogs are constantly offering up new perspectives and new techniques to make business process experts all around the world more effective.

"The BPX community is a place for frank discussion of your craft. You can determine what works well, what doesn't, and what is hard, without fear of being judged or attacked," said Axel Angeli. "You sharpen your skills, improve your arguments, and even engage in combat. You can find out from others how they solved problems you are facing. You get advice about how to persuade others to get on board."

Where Is the Business Process Expert Community Going?

The BPX community is maturing. The rapid growth and substantial collection of content that has been created has proven the point that the community and subsequent importance of the business process expert role is a growing part of creating effective solutions. Now that critical mass has been achieved, it is likely that the community will continue to evolve to more directly support the needs of business process experts. Here are a few of the directions that the BPX community could take:

- **Business Process Innovation and Product Development**: The BPX community could expand its scope of offerings to become the premier online community and marketplace for business process research, definition, modeling, application composition, and innovation (solutions and processes), on which community members can collaborate and share knowledge.
- **Marketplace**: The BPX community could become a primary place for co-innovation and product development, constantly where SAP customers and partners can model, market, demo, or sell solutions and related services.
- **Product and Solution Development**: The ability of the BPX community to reveal trends and user requirements could result in more communication between business process experts and Solution Managers who are designing the next generation of SAP products. It also could become one of the primary feedback channels for such development, for instance, via semi-private beta product and solution communities, where experts and customers can vote and react to features proposed. Imagine the possibilities to move tradi-

tional help documentation that comes with products to an online wiki environment where it can be kept up to date and improved by customers and partners as well as staff.
- **Increased Partner Presence**: As customer needs are increasingly met, not just by SAP products and services, but also by coordinated ecosystems involving partner offerings, partners may show up more prominently in the BPX community.
- **Events**: Someday, perhaps a global event or conference will take its natural place alongside SAPPHIRE and TechEd and provide a forum for meetings and education to promote the topics that make up the business process expert role and help everyone improve their craft.

While any of these developments would be exciting and useful, it is likely that the most important new directions of the BPX community will not be planned but will emerge from community members who see new ways for the community to serve their needs.

As noted at the beginning of the chapter, the BPX community continues to grow, aiming to support and enhance the work lives of the evolving resident business process experts. We want to remain the leading platform that enables you to connect with peers and share business process ideas, find answers for related business process and system implementation challenges, and find solutions for new or existing processes. The possibilities are indeed exciting. Imagine business process experts sitting in meetings with customers and finding answers on a mobile device in a second. Imagine members podcasting on the fly using their phones. Imagine members taking a picture of a whiteboard with a process idea and sharing that via a tweet with their closest friends to ask for feedback. Imagine private spaces where you are allowed to design on the fly from anywhere in the world. Imagine the ability to do VOIP video conferencing that is recorded on the fly and posted in the community for anyone to view or join. It is community meets business, accelerating solution creation by putting process first.

5 Organizational Change Management and the Business Process Expert

Organizations change because they want different business results, motivated by ambition as well as by pain. A company seeks to move to a new state of affairs because the management wants to spend less money, make more money, or some combination of the two.

Change is almost never easy. People rarely do things the way they do because they want to do a bad job. Current ways of working at most companies are the result of historical precedent, habits, reactions to past challenges, the influence of technology, and directions from senior management. Adopting a new way of working means leaving the world of the proven and familiar and entering a new world that at first seems strange.

Organizational Change Management is the field that is dedicated to managing change in an orderly and effective manner. While practitioners are able to point to a steady stream of successes and a growing body of best practices, even the most enthusiastic do not claim that change management as a discipline is an easy sell. "Change management is still something people have a better collective understanding of these days, but they don't entirely trust it," said Kerry Brown, Global Director of Organizational Change Management at SAP. "They know they need some of it, but they're not so sure what to do." Brown finds that people who have gone through disastrous attempts at transforming their organizations are the ones most willing to learn.

Broadly, change management involves two areas: strategic and organizational alignment. Organizational alignment involves coordinating the goals, expectations, messages, and priorities that will result in achieving new outcomes and is the highest level of change management. Once this is in place, the challenge in change management becomes tactical, project- or program-level change management. This involves making sure that everyone involved knows what they should be doing and when and how it is related to what everyone else is doing.

The reason that the field of change management exists is that people do not take easily to change. "There are four main reasons people resist change," said Brown. "I didn't know. I wasn't able. I wasn't involved. And, I'm not willing." Facing and conquering these forces requires communication, building of trust, hard and soft skills, and a combination of stubbornness and empathy. Success comes when people start to act differently, whether they like it or not, because they understand why they are being asked to change. It is only after the changes have stopped seeming strange and become habitual that the highest level of performance will result.

Put another way, change management is about the dimensions of people, process, and technology. The process defines the world you want to achieve. The people must be convinced to go there. The technology is the enabler. An understanding of change management is vital to business process experts because the task of addressing the issues related to successful transformation of an organization frequently have no owner and fall to them by default.

> **From the Wiki: Different Responses to Change**
>
> Change is the only constant in life, but people don't like to change. That is why it is important to introduce change gradually, in small increments, to maximize the chance of acceptance. Start with the easiest and simplest project that has the maximum chance of success, recommends Business Process Expert (BPX) community member Ashish Mehta. In every organization, roughly 10 to 15 % of people are innovators/early adopters who are willing to try something new while about 65% will go with the crowd (followers) and the remaining 15% or so are resistors, who chafe at any and all change, observes Mehta. To ensure success, you have to win over both the innovators/early adopters and the followers. The followers need a little convincing, which can usually be achieved with your first small success. As for resistors, you have to treat them with velvet gloves. Be careful not to let negativity spread, which is the forte of this community of people.

The Specific Challenges of the Business Process Expert

Business process change does not happen in isolation. Business process change becomes organizational change by definition, and in most companies, organizational change has no explicit owner. It is up to the team involved in effecting change to figure out how to manage the change and make sure that it sticks. This means that the de facto heads of change management are the project managers, line managers, senior executives, information workers, and IT staff, who are most committed to making the change happen. If the business process expert role is successful in even the slightest way, it means that business process experts become part of the change management team.

But, in a sense, business process experts are different from many of the other members of the team. Business process experts are outsiders. They do not own the business process they are changing. They are a coach or an assistant coach who provides advice. They are not on the playing field when the game starts.

Business process experts are also advocates and teachers of a general perspective on business, the business process perspective described in Chapter 1. This means that they are enabling the organization at large to embrace business process change and embrace the technology to perform differently and get different results. Business process experts cannot function in this role effectively without building trust and respecting the abilities of those who are on the playing field. They must be role models for an open mind and an eager attitude about change, guided by the business process perspective, without being overly prescriptive or exaggerating their own importance. It is a tricky path to navigate and in this section of the chapter, we examine some of the more difficult aspects of change management that business process experts face.

Promoting the Business Process Perspective

Whether a business process expert calls it "business process perspective" or not, one enduring change management challenge is advocating the central role of business processes in business transformation. The challenge in advocating for this perspective is that it is rarely possible to convince people in one attempt that the elements of the business process perspective are crucial to success.

In addition, parts of the business process perspective are aimed at senior management (commitment to adapt the organization to support business processes and work through challenges), other parts are more important to line managers (thinking in terms of the process, not the software, thinking in terms of the largest picture, incremental improvement, process documentation), and still others are vital to the technologists involved (thinking in terms of the process not the software, thinking in terms of the largest picture, incremental improvement).

In practice, this means that the business process expert must opportunistically advocate for the business process perspective, both explaining the theory and showing how various tactics that fit into the theory provide value. For example, if you create a document that describes the end-to-end process, circulate it to all involved, and take the time to get their reaction, many suggestions for improvements are usually harvested.

But the business process expert must walk a fine line between being an advocate for the business process perspective and being a nag or a pest. Each project is an opportunity to convince and educate but you don't have to win the entire battle and every heart and mind. The best advertisement for the business process perspective is to create a new way of working that delivers business results.

Business Process Maturity

The right role for a business process expert will change based on the level of maturity that an organization has with respect to the business process perspective and its ability to transform itself. Just introducing a business process expert into an environment that has never been friendly to change and has technology that is not easy to adapt is unlikely to succeed in the short term. The way that a business process expert will be most effective will vary based on an organization's readiness for the role.

If an organization is just starting out and there is little understanding of the business process perspective or the business process expert role, perhaps the best approach would be to only apply the new methods to a series of pilot projects in order to build up credibility and create more readiness in the wider organization based on a track record of success. Perhaps a part-time business process expert could be assigned to join these projects, someone from the project management team or another department with wide responsibilities.

If a company is highly methodical already and has a well-defined change management process in place and technology that is adaptable, then perhaps it makes sense to introduce a full-time business process expert who can participate in many different projects and add the missing elements of the business process perspective, such as the separations of process design from software implementation and a clear documentation of the end-to-end business process.

If a company is engaged in a constant cycle of incremental improvement with adaptable technology and is friendly to the adoption of the business process perspective, then perhaps it makes sense to have one member of each team trained as a business process expert so that they can bring the role to a wider audience.

Of course, these choices just scratch the surface of the different ways to adopt the business process expert role, but factoring in the existing maturity with respect to the business process perspective should be part of all of them.

Interaction with Existing Roles

While the business process expert role represents a new combination of responsibilities, many of the tasks included overlap with existing roles that are involved in planning for the future, process transformation, and creating solutions. In addition, the business process expert must ask and get commitment and support from other people in order to succeed. In other words, the business process expert is going to run into others in the organization who may see the new role as a threat or the support asked for as an inconvenience. If a business process expert does not handle these interactions with finesse and care, success will be quite difficult.

One of the implications of this overlap is that successful business process experts mostly seem to be resident inside a company, as opposed to outside consultants. It is one thing to have a new role show up on your doorstep that may overlap with your current responsibilities, but it's another for that person to come from outside the company. That may be too much to take.

Introducing the business process expert role into an organization must be done carefully in order to obtain the maximum chance for success. Business process experts do many things that other people in the company currently do now. Many people will rightly wonder: Am I being stripped of responsibilities? Others may see the arrival of business process experts as an opportunity to unload parts of their jobs that have been frustrating or difficult.

Perhaps the most important prerequisite for introduction of the business process expert is pain. Companies that are building effective and flexible solutions do not feel a need to change their way of creating solutions. It is only when a company is having trouble, or sees competitors doing better, that the motivation for change emerges. When needed solutions do not arrive, or when the solutions that do arrive do not meet expectations, the stage is set properly for a business process expert.

So far, the business process experts involved in our research are a sophisticated bunch. They all believe that the business process expert role must be accepted readily by those involved in solution creation for the highest productivity to ensue. Taking this acceptance for granted is a grave mistake. Attempting to dictate to a team that "you will work with a business process expert" is such a bad idea that it has never been attempted by anyone with whom we've spoken.

Gaining acceptance for the role is rarely easy and involves persuasion and education. Here are some of the points that business process experts have found are frequently raised with all of the different people involved in discussions about adopting the business process expert role.

- **Senior management** must be on board for a business process expert to succeed. In a perfect world, everything would go swimmingly on every project. But in the real world, at times, projects will bog down, success will appear far away, and some fingers will point toward the business process expert as the culprit. In times like these, it is important that senior management be supportive of the transformation toward the business process perspective. In selling the idea of the role to senior management, business process experts must be careful not to fall into the trap of setting unrealistically high expectations. The best approach is to explain the theory behind the role, the requirements for the people playing it, and the challenges involved in adoption, and then suggest a modest program of expanded experimentation and the introduction of metrics that will track progress. Often, the institution of a COO role or a Chief Process Officer role can help accelerate the buy-in from the lines of business, as well.

- **Line of business managers** see the business process expert as just another part of the process of creating IT solutions to automate their business processes. They aren't likely to have many objections about suggestions for improvements unless people on their staff complain that the business process expert role is a threat to success. If the business process expert role is successful, line of business managers will be happy but will not really care why it worked.

- **Process owners** own a process from the business point of view. They are the quarterback on the field, or perhaps the head coach who is responsible for success. For example, the director in charge of warehouse management in a large company may be looking at all the metrics about how the warehouse is run. The process owner wants the IT solutions to help implement processes that will help improve the business results. The process owners are frequently a fountain of information about areas that are ripe for improvement and innovation. The process owners care less about the details and more about the results. They can be powerful advocates for change if a business process expert can produce the results to gain their trust.
- **Process experts** lead the teams executing important parts of a process and using the software that keeps track of it. These are the people who know the ins and outs of the existing solutions and can point to specific shortcomings and strengths in the way things are automated. Business process experts tend to spend a huge amount of time with process experts, understanding the existing systems and brainstorming about improvements.
- **Business analysts** are focused on looking at the way a line of business works and providing analysis of business performance, processes, and technology. Business analysts are usually an important part of the innovation process and are expected to come up with suggestions for optimizations and improvements. The business process expert role can seem like a major incursion into this territory because of the leading role business process experts like to take in business process design and promotion of the business process perspective. In a company that uses the business analyst role, the business process expert must be carefully promoted so as not to cause conflict. In the end, business analysts tend to be fellow travelers with business process experts because they both are seeking to achieve the same goals. Over time, we expect a convergence of the business process expert role and the business analyst role. Perhaps it will not be necessary to create Visio diagrams anymore; instead, professionals can use modeling tools to make instant updates to

the systems that support the business process. We expect many business analysts to evolve into business process experts.
- **End users, also called business users**, use the systems. Business process experts work with end users to understand existing systems and to confirm that problems identified by process owners and process experts are the whole story.
- **Project managers**, like business process experts, frequently fill in gaps as needed in support of change management and communication. A business process expert must be sensitive to the existing role that a project manager is playing and how that fits in with the business process perspective. Frequently, project managers are among the easiest to convince of the value of starting the analysis of solution creation with the business process because they have seen things go wrong in so many other ways. If a project manager is playing a role that a business process expert expected to play, the right call is almost always to defer to the project manager. It doesn't matter who's doing what as long as the needed work gets done.
- **CIOs and CTOs** are the senior management level of the technology department. Business process experts sometimes have a challenge building trust with these executives because the imposition of the business process expert role can be seen as a failure on the part of the IT department. Political issues can arise when the topic of who business process experts should report to is discussed. But most of the time technology management understand the business process expert role and does everything they can to support it. Also, as the power of the CIO seems to be eroding, it is of interest to the CIOs of the future to heavily engage and support the creation of a business process expert organization.
- **Enterprise architects** are usually highly focused on business processes and find that the business process expert role can help raise the profile of BPM in the solution creation process. Business process experts frequently find that the most extensive descriptions of business processes and the most advanced modeling tools are used by enterprise architects. The fact that business process experts also attempt to exert leadership with respect to business process design and the planning for transformation of business processes over the long term can lead to tension with enterprise architects.

Much of the time this tension is resolved because business process experts are usually focused on the short and medium term and enterprise architects are primarily interested in the medium and long term. The two roles have powerful common ground in that both are interested in promoting the business process perspective. You can probably imagine the enterprise architect defining the IT landscape and the master business process expert (the business process architect) defining the process landscape.

- **Application experts** have a deep understanding of how to configure enterprise applications. Business process experts usually spend a lot of time with application experts understanding the way that processes are currently automated by the applications. Application experts also provide an understanding of what functionality is not being used and how the applications are service enabled. Business process experts can conflict with application experts over the right way to create a solution. Sometimes application experts think more in terms of the capabilities of the application instead of the ideal process, which is the perspective advocated by the business process expert. However, most of the time application experts and business process experts work in harmony to pursue implementation of processes that are most beneficial to the business. Similar to the business analyst, the application expert could easily evolve to become a resident business process expert.
- **Technology architects** are responsible for designing architecture for the software and hardware that implement solutions. Business process experts have long conversations to make sure that solution requirements are clear and that the business process perspective is central to the way solutions are designed. Business process experts can clash with technology architects if the needs of the business processes conflict with infrastructure choices, although such conflicts are rare.
- **Engineers and developers** build the solutions that business process experts help design. Business process experts spend time with developers understanding the effort involved in creating various aspects of the solution. Business process experts also get help from engineers and developers in evaluating the quality of model-driven development tools. Conflicts between business process experts and developers can arise if developers attempt to turn every problem

into a development project instead of using standard software as much as possible.

In each company the business process expert will fight a different battle to gain acceptance, prove the role is valuable, and then propagate it. While the nature of the battles fought at each company may vary widely, the cast of characters just described is usually the same.

Conflicting Perspectives

Some of the conflict mentioned in the previous analysis of existing roles in the enterprise is caused not only by overlapping responsibilities but also by conflicts between the business process perspective and other perspectives on how to create the best solutions for enterprise computing.

Owen Pettiford, an IT professional with 20 years' experience and founder of CompriseIT, has done some fascinating analysis on his blog of these other perspectives and how they conflict with the business process perspective.

Pettiford points out that the current collection of technology that surrounds enterprise applications offers a new paradigm for solution creation. It is now possible, through service-oriented architecture, application composition tools, business process management technology, and other mechanisms, to implement a process in a flexible manner and add anything extra that is needed by implementing a service. In other words, you rarely have to start from scratch. You also rarely have to live with, and be constrained by, what an application does, because so much is possible through configuration and composition.

However, unfortunately, the understanding of what is possible in the era of SOA and composition is sometimes a hard sell. Standing in the way are perspectives about how to build solutions that are stubbornly held, because at times they are profoundly right. Competing with the business process expert and the implied business process perspective are the views held by many other types of experts.

> **From the Wiki: Making Change a Win-Win Situation**
>
> "I have noticed an interesting thing during my time in IT," reports community member Ashish Mehta. "Effective change rarely results from top-down initiatives. That is where most consultants fail. They come into an organization with preconceived ideas and predefined ways of thinking and fixing problems. They push their product or solution and then encounter resistance from somewhere in the rank and file or sometimes even from management. Management buy-in is important, undoubtedly; if you do not have approval from management, you will not get very far in the project and funding could start to become a serious issue. However, even more important is rank and file acceptance. This is how I try to impart change. First, I talk with senior management, the head of IT or CIO as well as the head of finance or CFO to discover their main pain points. Then I mold my solution/product to meet their needs. I draw a picture of the workflow to illustrate what I plan to achieve. After that, I tell management that I want to hear what employees and end users are feeling. I spend time talking with users and figure out ways to help them do their jobs better. You know, most employees will gladly provide you with the information you need when you tell them you want to help them do their job better. In this way, resistance to change melts away, and the chances of success greatly increase. Finally I put the graphic of the process on the hallways and make it a visible issue for the entire company. It's all about creating a win-win situation."

- Application experts, who Pettiford calls platform application experts in his blog, tend to try to satisfy every business requirement with the capabilities of the existing application. While in the past this has saved time because it has forced some processes that were customized for no powerful reason into the standard form implemented in an application, it also has costs in terms of innovation. If a process that deviates from the idea is not supported by the application as it stands, a conflict can arise between people who want to implement the optimal process and those who want to change the process to meet the capabilities of the application. In their work, application experts and their kin say, "This is what the application can do; this is what you can have."

- Industry solution experts, partner solution experts, and composite application experts take the same perspective as platform application experts, except with regard to an industry or partner solution or composite application.
- Enterprise portal experts tend to see every problem as a chance to whip up a new portal application.
- Platform development experts and composite development experts see every problem as an invitation to custom development. They may be frustrated by the business process perspective because reuse increases and opportunities for new development are rarer and are focused on creating reusable services. At their worst, development experts say, "Let's make it perfect, just what you want, without regard to the cost now or in the future."

Pettiford does not claim that these perspectives are exhaustive. There are plenty of other narrow ways of thinking to add to these that constrain the world of the possible. The challenge faced by business process experts is to get everyone involved to rise above these perspectives when needed and adopt the business process perspective, which means designing the process first, and then implementing it as well as possible using enabling technology.

Lessons Learned about Adopting the Business Process Expert Role

Introducing the business process expert role successfully to a company is something of an art form. Each company is different with respect to its business process maturity, readiness for change, and motivation for adopting the role. While universally applicable advice is hard to come by, the business process experts we talked to in performing research for this book did have some insights to offer that may be helpful.

Grassroots Beginnings

Our research has shown that planning for a formal introduction of the business process expert role by a management-led team is almost never the way that business process experts first emerge at a company.

The most common path is that people start visiting SAP's BPX community site, or find out about the role in other ways on the Internet, and then they just start doing what other business process experts do. Most of the time, this results in delivering improved solutions, perhaps after a few false starts. But then, in most healthy companies, somebody asks the question, "Why did this project go well when others did not? What is different here?" That's when the discussion of introducing the business process expert at an institutional level begins.

At this point, the challenge becomes defining the role in a way that is likely to succeed at the company in question, to obtain the right people to become business process experts and give them the needed resources, and finally to make sure that the first wave of business process experts have the right skills and abilities. Then, if expectations are set properly, it is possible for the business process expert role to become accepted.

The Elements of Successful Adoption

While each of the companies that have adopted the business process expert role has gone on a unique journey, some common elements have emerged so far and a pattern of successful adoption is starting to take a general form:

- The **grassroots introduction** of the business process expert role happens when someone in the company thinks that using the role would make sense and starts to act as a business process expert.
- Early successes lead to a **discussion of wider adoption**.
- **Senior management buys into the business process perspective and the business process expert role** and decides to attempt to adopt the role more widely. This usually happens gradually. We have not found a companywide, big-bang introduction of the business process expert role.
- **Experimentation leads to refinement** of what the business process expert role should be for the company in question. Best practices are developed. The role is adapted to the business process maturity

and change management practices of the company. The required technology landscape is improved.
- **The business process expert role is institutionalized** and propagated. Formal job descriptions are developed along with training and certification. Business process experts start to show up on the organizational chart. Metrics and KPIs to track the impact of the role are developed and monitored.

The rest of the chapter touches on some of the thornier issues that arisewhen adopting the business process expert role.

Focus on Points of Pain

One way to maximize the return of applying the business process expert role is to focus the initial efforts on the worst performing parts of the company. Points of pain can be a trigger that spurs collaboration and action. On the other hand, this approach can be risky because areas that are points of pain usually have intractable issues related to personnel, politics, or shortcomings in the capabilities of technology. A business process expert must be brave to wade into such a situation. Even in these areas, change may be resisted. As Axel Angeli put it, "Why should I change anything? Everything works fine. The hammer falls on my toe once a day, but I've gotten used to it."

But fixing problems in a department that has been an enduring source of pain for a company is one way to build credibility that the role is worthwhile. If business process experts can help make the most negative situation better, it will be likely that they will be able to fix areas that are less pathological.

> **From the Wiki: To Change or Not to Change**
>
> Business users and project sponsors may rely on business process experts for recommendations about whether system changes are required or whether some workaround can be found. The business process expert uses his judgment to recommend changes or workarounds based on knowledge of user capabilities, cost implications, and possible benefits to the organization, according to BPX community member Ajay Ganpat Chavan.

Focus on Strategic Projects

Another choice that is made frequently is to apply the business process expert to the most important solution creation project facing a company as a form of insurance that things will go well. Making this choice means that the stakes are high for the business process expert. If anything goes wrong, they will likely share the blame. But when management introduces the business process expert role into key projects, it is also a vote of confidence in the role and the business process experts who are assigned.

Focus on Enthusiasm

Following the enthusiasm of the staff for adopting the business process expert role can be a low-risk way to create demonstrations of how the role can help. Under this scenario, management explains the new role and then allows teams involved in solution creation to introduce the role by accepting a new member or by sending one of their team for training. In this way, the company takes the path of least resistance and promotes the business process expert role where it is wanted most.

Create a Center of Excellence

Creating a central team of business process experts is yet another way of introducing the role. Under this scenario, a central team of business process experts that can be assigned to projects as needed is created and trained. This method frequently works well because the central team acts as a clearinghouse for information about how to successfully adopt the role.

Metrics and KPIs for Tracking Success

Business process experts frequently make a difference in many ways, both tangible and intangible. Metrics and KPIs for business process experts are not a luxury or something that is helpful or nice to have; they are a matter of survival. But senior management rarely is impressed by enthusiastic testimonials—they want demonstrable results.

Without metrics, the business process expert role is something you have to have faith in. This is not a reasonable basis for commitment to the business process expert role. There must be some kind of success metrics so that the value can be documented and the business process expert role can be justified during inevitable downturns when organizations shrink.

While it can be a challenge to construct metrics that show the full value of the business process expert role, showing impact in the following metrics has proven helpful in making the case:

- Reduction in cancelled or failed projects
- Reduction in length of projects
- Increase in projects delivered on time
- Increase in user suggestions harvested and implemented
- Increase in efficiency based on better processes and solutions
- Increase in clock speed for the cycle of incremental improvements
- Increase in number of users getting the benefit of existing technology

While some of these metrics are pretty soft and fuzzy and will be hard to track, it is vital that persistent effort be made by business process experts to demonstrate the value of their role and of the business process perspective until the entire organization has been shown the benefits.

Mistakes to Watch Out For

Becoming a business process expert is a truly satisfying endeavor. The sheer pleasure of mastering the skills and the feeling of really making a difference in a company is evident in the continuing waves of enthusiasm that are expressed on the business process expert community. But few who have mastered the role report that it felt easy.

One of the things that the research for this book revealed is a collection of pitfalls that can retard the progress of a business process expert. We list them here in the hopes that readers will be inspired to add to this list by joining in the discussion at the wiki version of this book.

Intolerance for Ambiguity

It is a mistake to expect that things will proceed in a highly structured predictable manner in your work as a business process expert.

To succeed, Kerry Brown suggests that business process experts must have a high tolerance for ambiguity, meaning that they must be comfortable with the many distractions that will arise and the fact that there may be many ways to get to the end result.

"If you look at the business process expert role, they are basically living constantly in a project lifecycle, if you will," said Brown. "They have to be agile and adaptive to new information, new data, new requirements all the time. And so, being able to say, 'I'm going to go work on this from A to F, and I know that these are the six steps in between,' is naïve, because they say they're going to work from A to F, but they're going to have 262 distractions between A and F."

A Sense of Ownership

It is a mistake to act as if you are an owner of the processes you are helping improve.

Business process experts help transform the processes and supporting systems at a company, but for the change to be complete, the team that owns the process must feel that they were responsible for the change and have a stake in making the new way of working successful.

"The business process expert role is by definition a facilitator, consultative role, collaborative role, so business process experts don't ever actually own the outcome," said Brown. "They own contributing to the outcome and accelerating the outcome, but they don't own the outcome. They don't own the process. And so they've got to take a consultative approach, whether or not they're internal, external or otherwise, in order to influence people to change, versus attempting to control that change. Sustainable change comes from creating shared ownership."

> **From the Wiki: Pass It On**
>
> The solution creation that business process experts help facilitate is the product of true collaboration among various parties, so the business process expert shouldn't be seen as a miracle worker with a halo. It's true that the language used by IT professionals frequently distances them from the business units, says BPX community member Anbazhagan Sam Venkatesan. Business process experts come on the scene as facilitators. The solutions that they help facilitate are the outcome of a collaboration of the departments representing users (sometimes called the user department), the IT department, and business process experts.
>
> Any solution requires implementation, in which the user department plays a key role, with intermittent interaction with the IT department and perhaps occasional interaction with the business process expert. A successful business process expert passes on the ownership of the solution to the user department and IT.

Not Listening—Thinking You Are the Expert

It is a mistake to attempt to teach first and listen second.

Business process experts are usually proud of what they know and are eager teachers. But the time for teaching is usually after the people who own the process have had a chance to speak. The people who own the process, who are toiling with it every day, usually have a great perspective about what is wrong and what could be better. This should be the foundation, and the knowledge and skills of the business process expert should come out opportunistically, as needed.

Overpromising

It is a mistake to set expectations too high.

The world of technology is full of people who are so eager to help that they downplay difficulties and assume that the world moves faster than it actually does. Business process experts, like all technologists, can fall prey to this error. Seeming too optimistic can reduce credibility with people who are experienced and know how long things take and how difficult change can be.

Staying Hands On

It is a mistake to take too much of an advisory role.

Business process experts are not just consultants; they must join in projects and help do the work. If you tell a team to try something and report back on how it went, you will get a much different reaction than if you join in the process of change and help as you can.

6 The Technology Environment for the Business Process Expert

In the strictest sense, the business process expert role should be technology-agnostic, never promoting technology for its own sake or favoring one form over another. A business process expert should be able to help a company do better even in a situation in which no technology is brought to bear, simply by promoting the business process perspective.

But in the modern world of business, everything is intermediated and supported by technology. Discussions of the best practices for business processes quickly lead to the discussion of solutions that support those processes. Technology plays an important role and the business process expert must be fluent in helping organizations find the right tools for the right job. After all, not all technology for collaboration is created equal; some collaboration platforms work better than others to support working in teams, for example.

The challenge for this chapter is that the world of technology is so vast and changes so quickly that it is hard to summarize how the business process expert can best make use of it all. The need to effectively apply technology is one of the reasons for the creation of the SAP

Business Process Expert (BPX) community, which helps to assist business process experts through the ins and outs of the technology maze.

Experience has shown that business process experts must be armed with technology and supporting capabilities to make the maximum contribution to their companies. Once a company has adopted the business process perspective, for the business process expert to be most effective, companies must have a minimum of business process maturity and technology infrastructure. Companies that have cultivated and nurtured their technology landscape to meet the needs of their processes will be the first to benefit from a business process expert's talents.

Many business process experts make a difference right away. It is possible to make quite an impact through informal means that lead to improved communication in the solution creation process. But the first wave of business process experts have found that larger and longer lasting benefits are created by institutionalizing communication and expanding participation in solution creation with supporting technologies.

The ideal technology landscape is rich in functionality and offers substantial flexibility. If all roads lead to custom development, a bottleneck will be sure to develop. If all roads lead to an ERP system, then the scope of configuration of ERP will become the constraint. If it is hard to communicate and collaborate, solution creation will be slowed down. If no emphasis is placed on defining solutions at the BPM level, then an application-centric view of the world will fill the void. If the infrastructure for SOA is immature and unreliable, it will be hard to create flexible solutions and react quickly to changing process needs.

Today we believe that the preferred technological environment for a business process expert is one in which a strong foundation of configurable SOA-enabled enterprise applications is supplemented by a dial-tone for collaboration and the ability to use services and business process modeling tools to accelerate the creation and implementation of composite applications and mashups. Availability of tools for modeling and development can also expand the number of people who can participate in solution creation. The same is true for simplified ways of configuring applications or user-friendly methods of composing new solutions based on services. An environment with all these elements provides a rich pallet for automation.

The question that this chapter seeks to answer is simple:

What are the enabling technologies that each category of business process expert needs to do his or her job to maximum effect?

In Chapter 3, we described five styles of business process expert, each of which puts the empathy, technology, and change management skills of the business process expert to work in a different way. The type of role that may be appropriate differs for each organization. In some organizations, business process experts may play one or more of these roles or combine them in new ways. We are still in the early days and new discoveries happen at every company that adopts the role.

This chapter looks at the technology that is related to each role a business process expert may play. Of course, all technology potentially applies to any role that a business process expert may play. We are grouping the technologies discussed by role not as a strict categorization but as a way to organize our discussion of this vast topic.

Technology Related to the Business Process Expert as Organizational Therapist

In the organizational therapist role, the business process expert identifies and solves communication problems. In environments with poor communication, conflicts become routine and mistrust grows. Abusive and disrespectful behavior becomes a habit.

Technology does not directly play a role in resolving this acrimony. This is the task of a business process expert who gains the trust of both sides and reestablishes trust that paves the way for communication and productivity.

But once a business process expert has treated the initial pathology and the technology and business sides want to communicate with each other and work together, establishing an easy-to-use dial tone for communication and collaboration can be an investment that pays huge dividends. Business process experts who act as organizational therapists help recreate communities that have broken down or never existed. Technology can play a role in sustaining those communities.

A company introducing the business process expert role will find that the technologies mentioned here will help amplify the impact of a talented business process expert and institutionalize positive patterns of behavior.

Video, Blogs, and Podcasts

We often think of blogs as a global megaphone that allows one person to broadcast her thoughts. But blogs, whether written, video, or audio, can also be used as a collective communication mechanism for a group of people working on a project. In this scenario, the blog is not just a forum for one person but can contain entries from anyone working on the project. In this way, thoughts and questions can be captured and the entire team becomes aware of each other's thinking. The best blogs are often a summary of what the business process expert experienced that day, week, or month and are used as a collection of useful links by others that have limited time. Everyone can also participate in answering questions.

Project blogs have proven to be an effective way of proposing, refining, and propagating best practices.

Wikis

Wikis create a shared space to support collaboration and shared content creation, which can be used for project management, requirements gathering, creating documentation, and many other activities.

Wikis often take root in unpredictable ways. Frequently the best approach is to introduce them and use them for a particular purpose and then a community of users forms and takes the wiki in many different directions. As wikis evolve to become more tailored to the business, we will see many different types of wikis, including process wikis, content wikis, spreadsheet wikis, and more.

Instant Messaging and Group Chat

One of the lessons of agile development is that developers must be in close proximity with the users for whom they are creating applications. Some practitioners of eXtreme Programming suggest that the development team be within 20 feet of the users so that questions can be asked and answered quickly.

Another way to achieve this intimacy is through instant messaging. If developers and users can ping each other with an instant message to answer questions quickly, then the flow of development can move quickly forward. Some companies, like Yahoo, have already moved their primary way of communication from email to IM.

Teleconferencing, Webconferencing, and Videoconferencing

When blogs, wikis, and instant messaging are in use, frequently important issues bubble up that require input from more people. The ability to bring teams together quickly through (VOIP) teleconferencing, webconferencing, and videoconferencing often accelerates communication.

Ease of use is the key to achieving benefits from these collaborative technologies. The faster the path from thought to collaboration through one of these mechanisms, the more likely they will be used. If it is easy to go from a teleconference to a web conference for screen sharing, that option will be used frequently.

Information on Demand

To have the ability not only to broadcast, but also to replay information on demand for an end user, offers the possibility of assimilating new ideas, processes, or solutions at their own pace. This is increasingly important as most organizations are becoming globally flat.

Technology Related to the Business Process Expert as Requirements and Process Analyst

The business process expert who acts as a requirements and process analyst uses tools to capture knowledge and represent it in forms that describe solutions. The collaborative technologies just covered can play an important role in capturing requirements and improving them through collaboration.

Much pioneering work is currently being done in the realm of business process management (BPM), the practice of designing solutions by modeling business processes in abstract ways and then using those models to create solutions.

> **From the Wiki: The Importance of Secure Coding Skills**
>
> Business process experts need to have a sufficient understanding of security in general, and secure coding in particular, states community member Ashish Mehta. Such knowledge is necessary to ensure that all or most software developers in the company are trained in secure coding skills. Security is an important requirement to include in procurement specs for third-party tools or software purchased from contractors.
>
> Even if this tough stance on security slows down the application development cycle and makes it harder to meet deadlines, it should not matter. Secure coding practices ensure that the company avoids undue embarrassment from security flaws and breaches that could inadvertently cause massive loss of reputation and customer goodwill. Further, as more and more business process experts insist on secure coding practices, software vendors increasingly work toward hardwiring security into software.

Categories of Modeling Technologies

The BPM space is not easy to understand because there is a huge amount of activity at all levels. For a business process expert, it can be a daunting task to determine which of the available BPM technologies will actually be most helpful in achieving their goals. Only the largest companies have a hope of actually productively employing all the BPM technologies available. The following discussion explains the different categories of BPM technologies and suggests ways that business process experts may put them to use.

High-level Business Process Modeling

The highest level of BPM is one in which the components being modeled are abstractions that contain a large amount of functionality described in simplified form. This sort of modeling can be performed when defining an enterprise architecture for a large company or mapping out the structure of a large application or a process that flows across many applications. An example could be the ARIS environment created by IDS Scheer that allows modeling of business process components and the relationships between them. These models can then be used to control the configuration of enterprise applications, or as communication devices, or for other forms of automated solution creation. ARIS is currently used for SAP as well as Oracle systems.

Business process experts can use this form of modeling when working in concert with enterprise architects or application experts in designing end-to-end business processes. However, the downside is that these types of tools do not allow for instantaneous deployment.

Detailed Business Process Modeling

While high level modeling creates a map of the process components in use at a business, detailed business process modeling maps out the details of how a process is implemented. This modeling is still at a higher level than the logic used for implementation, but at this level reusable services that are a part of service-oriented architecture (SOA) start to appear as units of functionality that are invoked by elements of the model. We will expand our discussion of SOA and its relation to the business process expert in "Technology Related to the Business Process Expert as Solution Designer," later in this chapter.

Just a couple of years ago, there were many proprietary products at this level of modeling, but a groundswell of momentum has established Business Process Modeling Notation (BPMN) as a de facto standard.

Business process experts use this sort of modeling to accelerate the design of solutions and to map out requirements in a detailed, unambiguous way.

> **From the Wiki: The Importance of BPM Security**
>
> Increasingly, business processes are moving outside the enterprise through the mobility of applications on laptops, PDAs, and Blackberries, notes community member Ashish Mehta.
>
> The business process expert must stay on top of this trend and help to coordinate and organize the use of mobile technologies. With the increase in oil prices, the need for telecommuting, web-conferencing, and working from home is on the rise. It is important that all the technology tools business process experts use are flexible and secure enough to accommodate this trend.
>
> A suitable enterprisewide encryption policy must be implemented in accordance with requisite compliance requirements, whether Sarbanes-Oxley (SOX), Payment Card Initiative (PCI), the Health Insurance Portability and Accountability Act (HIPAA), or others. Business process experts need to work with IT security officers, IT administrators, and the CIO. Because business process modeling is so strategic, encryption of each part of the cycle is essential. Think what would happen if the business workflow process ended up in the hands of a competitor because of a stolen laptop. In particular, workers must not turn off the encryption on their laptops in order to speed them up, favoring convenience over security.

Software-Focused Modeling

Many of the models at the high level and detailed level end up being used to create software, either automatically or through some manual process of translation. But a large amount of effort has gone into modeling the software itself, usually using the Unified Modeling Language (UML), an object-oriented approach to describing the structure of software. UML is a rich modeling environment that has had significant influence on subsequent generations of modeling technology.

Business process experts may find that UML is an appropriate way to communicate with developers about the structure and requirements for solutions.

Model-driven Development

Model-driven development is the practice of creating software using modeling as the way to describe the structure and behavior of the solution. All of the levels of modeling discussed so far are used by different projects. Business Process Execution Language (BPEL) is a language for describing business process behavior based on web services. Using BPEL, you can map out the invocations of a series of web services and how the results of each web service will be used to invoke the next one.

Model-driven development is clearly a part of BPM but it is more relevant to the role of business process expert as solution designer and we will discuss its use by business process experts in "Technology Related to the Business Process Expert as Solution Designer" later in this chapter.

Crawling, Walking, Running toward BPM

Exploiting BPM technology as a business process expert or in any other role is not something that happens quickly or easily. A company will not change from a mindset focused on applications and transactions to one focused on business processes overnight.

David Frankel suggests that one approach to adoption of BPM technologies is to think of a progression from crawling to walking to running. Each company will have to come up with its progression based on its own starting point and the existing technologies and skills it supports. The following definition is just an example that suggests how a business process expert might see the transition toward BPM adoption.

Crawling: Adopting a BPM mindset

For most companies, crawling toward BPM maturity means that when creating solutions, people start to think of the business process first and then ask how it can be supported. This is a significantly different way of thinking compared with the application-oriented mindset at many companies.

When process descriptions are increasingly used as the basis of requirements gathering and solution design, the company has started to crawl toward BPM.

Walking: Connecting Services and BPM Technologies

A company can be considered to be walking toward BPM maturity when it has started to create solutions based on model-driven techniques and the entire enterprise architecture of the company is described using BPM descriptions.

For this to happen, a company must also have significant maturity in terms of SOA. BPM models must have services to invoke in order to record transactions and make things happen in the world of enterprise applications. Services also are used to invoke external utility functions and to interact with partners. For BPM to be effective, a significant number of services must be available either through service-enabled enterprise applications, homegrown development, or third parties.

Running: Model-driven Process Automation

A company can be said to be running in terms of BPM maturity when the solutions are controlled primarily using BPM mechanisms. Application configuration, coding of services, and other techniques will always be in use, but, at some point the bulk of the work in a company will shift from development to modeling. It is at this point that a company can be said to be off and running in terms of BPM maturity.

Technology Related to the Business Process Expert as Solution Designer

Business process experts who play the role of solution designer work under the hood of an application, making sure that the process is automated in a way that balances the needs of the business process, the functionality of the application, and the technical capabilities of the organization. In this role, the business process expert uses empathy for the needs of the business, combined with an intimate knowledge of the available development and software configuration tools, to make whatever tradeoffs are needed to create a flexible solution for the lowest cost.

The ideal technology landscape for a business process expert playing the role of solution designer is one that is rich in services, application functionality, and development tools, so that the tradeoffs and compromises are as small as possible.

SOA Infrastructure

Increasing the maturity of the SOA infrastructure in the technology landscape is perhaps the single most important way to help empower business process experts to make a positive contribution to a company. The journey to SOA maturity is not something that can happen overnight, and the level of infrastructure needed will vary from company to company. Here are a few elements of SOA infrastructure maturity that will assist business process experts in doing their jobs:

- **Service-enabled applications** provide the raw materials so that BPM tools can create solutions that use and transform information in systems of record
- **A services repository** provides a directory where services can be found along with descriptions of their interfaces, functionality, and usage characteristics
- **Service development tools** allow custom and composite services to be created and added to the repository
- **SOA governance** provides a framework for deciding which new services to create, which service-enabled applications to deploy, and how people who use services can sign up to use them in the context of service-level agreements that describe the responsibilities of service providers and consumers
- **An SOA operational environment** provides the tools needed to manage the provisioning, operational monitoring, and dependencies between services that are deployed to support applications.
- **SOA training** teaches the skills needed to promote adoption
- **Enterprise architecture** provides the big picture view that makes sense of which services are provided by whom and how they should be used to support core business processes

The more of the infrastructure just described that is in place, the more raw materials a business process expert has to work with.

Model-driven Application Configuration Tools

Enterprise applications and other software solutions are increasingly adopting BPM methods in their design and configuration. These methods allow end-to-end processes to be specified using high-level modeling tools like ARIS. The high-level process descriptions are then used to guide the configuration of the applications to implement the end-to-end processes and also allow the applications to be monitored at the level of business transactions, not just in terms of software metrics such as disk space, memory, and database transactions.

Model-driven Development Tools

Business process experts thrive when solution creation is accelerated by model-driven tools. These tools speed the creation of solutions and emphasize the nature of the business process as the central focus. Model-driven tools simplify development by allowing those creating the solution to express the relationships that define the solution using higher level simplified abstractions.

Model-driven tools allow business process experts to create solutions faster, to teach others to build solutions, and to adapt solutions more rapidly to changing business conditions. Because creating and adapting solutions is easier, business process experts can support agile development methods based on rapid iterations.

The universe of model-driven development tools covers a lot of ground and can include any of the following technologies:

- **User-interface widget frameworks** allow users to assemble dashboards that bring together information from many services. As these frameworks mature, they are allowing more complicated relationships and interactivity to be expressed between the widgets using event mechanisms
- **Mashup and composition environments** allow users to assemble more complicated solutions from collections of services. In the simpler mashup environments, it is not possible to express much application logic, but the more advanced composition environments allow complicated logic to be expressed to process the results of service invocations

- **BPM application development environments** allow applications to be defined in term of high-level business process abstractions. Frequently, BPMN or BPEL is used to describe business processes.
- **Visual development environments** are model-driven development environments that emphasize graphical description of solutions

Partner Solutions and Services Marketplaces

The landscape for functionality for almost every domain of enterprise computing contains offerings from independent software vendors and services offered over the Internet in various forms of cloud computing such as software-as-a-service and platform-as-a-service. These solutions can play key roles in supporting and automating processes and can be incorporated by embedding the functionality in existing user interfaces, through web services, or using other forms of integration, such as Adobe Interactive Forms.

These categories are far from comprehensive and overlap to a certain degree. Some development environments combine aspects of all the categories mentioned. It is unlikely that even the largest companies will employ all of these tools. The main point for business process experts is that a company that wisely chooses tools that fit in with its application strategy will find that they are more productive in promoting creation of effective solutions.

Technology Related to the Business Process Expert as Empowerment Coach

Business process experts who play the role of empowerment coach seek to help people do things for themselves. For business staff, this may mean learning how to develop solutions using model-driven development tools. For technical staff, empowerment may mean learning to use collaborative tools to promote communication between team members.

The technology landscape for the business process expert as empowerment coach can include any of the technologies used so far. However, it is believed that the use of community and collective collaboration tools such as wikis, forums, blogs, and video will accelerate the notion of empowerment.

The empowerment coach role encompasses all the other roles, encouraging the use of the right technology to enable users to create, or aid in creating, solutions that meet their needs, as well as identifying the key evangelists that can carry the business process imperative and solutions forward in the realms of their ecosystem.

Technology Related to the Business Process Expert as Innovator

Business process experts who act as innovators need to work with many different kinds of people using various collaborative techniques. When a business process expert has a primary responsibility for innovation, he may need a way to develop ideas from early stages to a more mature form by bringing in a multi-disciplinary team. In such cases, wikis provide a valuable environment, as do blogs and forums, either publicly available or with restricted access. Using these tools, an idea can be taken from a seed into a more mature form that is ready for wider evaluation.

Many companies are finding that social networking is a valuable tool for people with like interests to come together for collaboration, knowledge sharing, and brainstorming. This general technique can be use to organically assemble teams for innovating in particular areas. Social networking voting mechanisms and environments such as research marketplaces can be used to evaluate proposals for innovation, supplementing the evaluation process with a form of the wisdom of crowds.

7 Patterns of Success

Enough experience has been gained by people who have worked as business process experts at companies throughout the world that patterns of success and failure are starting to emerge. Some of these patterns are proven while others are more speculative, but all of them come from the experience of business process experts.

Pattern is a word that is used in many senses in the world of software and technology. The meaning that we want to use in this chapter is the idea of a pattern of behavior or the application of technology that helps business process experts achieve repeatable successes. Patterns could be as simple as the idea of using a blog to keep track of project management information and a wiki to record requirements or create associated help documentation. Another pattern might be the suggestion to use graphical business process modeling notation (BPMN), also called swim lanes, to capture process descriptions during requirements gathering, even though you have no intention of using BPMN for automating solution creation.

This chapter is intended to be a repository for patterns of behavior and application of technology that have proven helpful to business process experts. It will always be a work in progress. In its current form, this exists as a request for content to be provided by the business process expert community (BPX community for short). Please feel free to add anything in this chapter that helps achieve repeatable success by visiting the wiki for this book. Simply google "Your Community BPX book" or visit:

https://www.sdn.sap.com/irj/sdn/wiki?path=/display/SEB/Main

The following headings and questions are only suggestions for ways to get started. The authors hope that members of the community feel inspired to add their experience to the wiki and move this chapter in any direction that may be helpful. The authors would like to thank Paul Taylor, from Kraft, for adding the first two patterns of success, included in the next section.

Patterns That Promote Adoption of the Business Process Expert Role

Our research so far has revealed that the most common pattern of adoption of the business process expert role involves an individual being inspired, then learning more about the role through the business process expert community, then performing the role at a company, which generally leads to wider recognition of its value and some form of institutional adoption. This is just one pattern however. Paul Taylor mentions two other patterns that lay the groundwork for successful adoption of the business process expert role: enterprise architecture and business process management (BPM).

When looking at the key patterns that determine the successful introduction of the business process expert role to organizations, a couple clearly stand out.

Enterprise Architecture

A formal approach to Enterprise Architecture (EA) helps mandate an alignment between key business and IT strategies. Not only will this prepare the ground for finding common agreement on today's requirements, but more importantly, it outlines the IT capabilities needed for the organization to be successful in the future. This high-level alignment will in turn provide support and help sustain a fledgling business process expert biosphere within an organization.

The EA style in question is not the main point here. It does not really matter whether an organization has deployed The Open Group Architecture Framework (TOGAF), the Zachman Framework, or another EA framework such as the SAP Reference Architecture. What matters is that the business process expert has a clear starting base when looking at the business process requirements and capabilities that need to be deployed in the future, and that they can be reasonably sure of the roadmap required to get there. This gives them a head start on trying to identify the areas that need to be addressed first and potentially the toolsets they will be able to use.

One other key element and outcome of this is the formation of architecture "domains" that focus on distinct business process (and technology) areas where both IT and businesspeople mingle together. This not only helps to break down barriers between business and IT, but also helps identify people who have the potential to take on a business process expert role in the future.

BPM

A formal approach to Business Process Management (BPM) helps the business process expert find traction within an organization in a number of ways. Such a framework provides a consistent approach to business process design as well as a means to monitor business processes. Thus armed with this approach, the business process expert can build on a solid foundation that provides a common way to describe business processes:

- The BPM notation to be used is agreed upon by business and IT, whether in the form of BPMN or Enterprise Process Center (EPC), and a consistent approach is used
- A method for business process execution has already been agreed upon by using either core ERP systems or leveraging composite applications across a composition environment such as that found in SAP NetWeaver 7.1 or both

- Approaches to business process design have already been codified within an organization and there are likely to be appropriate business process owners with whom the business process expert can work
- Business process governance channels have already been set up and approval paths identified, and thus the key design elements and decisions can be readily approved or escalated as appropriate

An agreed-upon business process improvement framework should have been established with clear requirements regarding measuring and justifying business process improvements. This allows the business process expert to confidently present the right types of improvements, ones that show the highest returns to the organization for any given scenario and that are consistent with previous business cases presented.

Identifying Additional Patterns

Here is a set of questions that may help identify additional patterns that may be useful in accelerating adoption of the business process expert role:

- How was the business process expert role adopted at your company?
- What barriers stood in the way of initial adoption?
- What barriers stood in the way of widespread adoption?
- How did you overcome those barriers?
- What methods proved effective for training people in the business process expert role?
- What methods proved effective for explaining the role to a wider audience?
- How has the practice of the business process experts at your company been improved based on experience?

Patterns That Promote Creation of Successful Solutions

Business process experts play a central role in the solution creation process, advocating for the business process perspective and a "process first" mentality. Business process experts report that they provide assistance clarifying the business requirements and the expression of those requirements in a definition of an ideal process. Using their empathy for business requirements, and their ability to translate that vision into a form that the technology department can understand, is another function that business process experts often carry out. Business process experts are also involved in the design of solutions, working closely with enterprise architects and software developers.

Here is a set of questions that may help identify patterns useful in creating effective solutions:
- How do you describe the ideal end-to-end process and communicate it to the team?
- How do you solicit comments on the proposed process and incorporate suggestions from all stakeholders?
- How do you confirm that the ideal process is in fact ideal?
- How do you use iterative development methods?
- How do you map the ideal process to the available technology?
- How do you decide when to compromise on implementing the ideal process and use existing functionality instead?
- How do you justify the cost of custom development to implement parts of the ideal process?
- How do you work with the technology team to design solutions?

Patterns of Communication and Collaboration

Frequently, business process experts arrive in the wake of one or more failures to create solutions that were urgently needed. While, in most cases, there are many things going wrong at once, poor communication is almost always one of the causes along with limited support for needed collaboration. One of the most difficult and effective things that a business process expert can do is to build trust so that communication is restored and then to teach new methods of working together to keep that communication flowing.

Here is a set of questions that may help identify patterns of communication and collaboration:

- How do you determine what went wrong with communication and collaboration on failed projects?
- How do you build trust so that people are motivated to communicate?
- How do you restore enthusiasm in the wake of failed projects?
- How do you introduce support for collaboration?
- How do you document processes so they can be widely understood?
- What patterns of meetings keep people on the same page?
- What project management methods tend to help restore effective communication and collaboration?

Patterns of Organizational Structure

The business process expert role does not have a natural home. Some organizations have business process experts reporting to IT, others have them reporting to the office of the CFO or COO, while still others have them reporting to lines of business. Business process experts are sometimes full-time employees who work on many projects across lines of business and other times are people selected from lines of business who spend part of their time working as business process experts. Organizational structure for the business process expert is highly dependent on the characteristics of each individual business.

Here is a set of questions that may help identify patterns of successful organizational structure for the business process expert role:

- To whom do business process experts report?
- Is there a center of excellence and an internal community for business process experts?
- How is the value provided by business process experts measured?
- Are there variations in the roles played by different business process experts?

Afterword: What We Learned about Writing a BPX Community Book

Now that the BPX community book project has reached another milestone, the completion of the second version, it is time to take stock and share what we have learned and how to improve the next time, as well as show leadership to other wiki book initiatives that might arise from this project.

The key takeaway is that it appears possible to combine the skills of professional writers with the emergent energy of community members, whose enthusiasm is directly related to their sense of ownership in guiding the shape of the content and making sure it serves the needs of other community members.

The key tension that was reconciled in the hybrid process was between the need for individual virtuosity that is the source of great writing and compelling narratives and the need to preserve enthusiasm from the community, which must be able to push the content in any direction that it feels must be followed. To completely understand what we have learned, the story of how this book came to be must be told.

The Collaborative Process

When the BPX community team asked the professional writing team at Evolved Media to join in the creation of a book about the business process expert role, everyone was extremely excited by the challenge. Evolved Media had created 10 books, including two book-length projects published on wikis (SAP Business One To Go and the Enterprise Service Bundles documentation). So, the idea of creating a book published on a wiki using a process that combined professional skills with contributions from members of the BPX community was an appealing idea.

Everyone involved brought a wealth of experience to the project. Marco ten Vaanholt, Global Head of the SAP Business Process Expert community, was present at the creation of the community and has an extensive network of experts with a wide variety of experience in making the business process expert role work effectively. (See the Preface section for a list of everyone who contributed.) Dan Woods, CTO and Editor of Evolved Media, had written *Wikis for Dummies* and had used wikis for more than 9 years in various ways. Dan has been studying the way that people use SAP software for more than 5 years. The most valuable asset is the BPX community, hundreds of thousands of people who are making the role work in all sorts of companies.

The process used to create the first two versions of the book was a modified version of the Communication by Design methodology that Evolved Media uses to write its books and that was heavily influenced by the idea of the Tom Sawyer wiki.

Communication by Design

Evolved Media writes books using a method called Communication by Design that combines aspects of journalistic practice with software development methodology. The method works through a division of labor that involves the following roles:

The **editor/analyst** is the equivalent of a software architect. This person designs the content, performs the research interviews (all of which are transcribed), and writes the most important parts.

Content owners are the sponsors of the project who work with the editor/analyst to make sure that the content being created fulfills the intended mission as described in a content requirements document.

Subject matter experts are people with knowledge and experience about the context for the content and the concepts and arguments that will be explained.

Writers are people who take the content requirements, the content design, the transcribed interviews, and other research materials and write first drafts.

Copyeditors, transcriptionists, graphic designers, production editors, and various other staff all play key roles in creating the content.

The content creation process takes place through each of these roles participating in the following steps:

Content requirements: The content owners and editor/analyst document the mission of the content and the requirements by which success will be judged.

Initial research and content design: A series of initial interviews by the editor/analyst with subject matter experts helps determine a content design, an outline of content that will meet the requirements.

Content design review and research plan: The content owners review the content design with the editor/analyst, improvements are made, and then subject matter experts are assigned to each part of the content.

(continued page 120)

> **Research**: The editor/analyst interviews the subject matter experts about specific portions of the content design. The interviews are transcribed. Other research materials are collected and associated with specific portions of the content design.
>
> **First draft**: The editor/analyst or writers takes all interviews and research associated with a specific portion of the content design and writes a first draft that is reviewed and improved by the editor/analyst.
>
> **Review**: The first draft is reviewed by the content owners and other subject matter experts, who make comments and suggest improvements.
>
> **Second and subsequent drafts**: Further research is performed and refinements are made to the content design as needed. Second drafts and subsequent are then created by the writers, which are reviewed by the content owners and subject matter experts.
>
> This process allows the knowledge of numerous subject matter experts to be harvested and focused on creating content with a specific mission. Each role in the process performs the task best suited to their skills. Evolved Media's books are routinely based on more than 100 interviews with subject matter experts.

Starting in February 2008, the Communication by Design process was used to create a first draft based on interviews with subject matter experts suggested by Marco ten Vaanholt. The content design and first drafts were posted on the BPX community wiki in April where community members weighed in with comments and additional content.

Avoiding the Tom Sawyer Wiki

This process avoided a phenomenon Dan Woods calls the "Tom Sawyer wiki" (see http://www.evolvedtechnologist.com/section/stories/avoiding-tom-sawyer-wiki). Such wikis are attempts to create content in which someone puts up an empty wiki, perhaps with pages dedicated to specific concepts or questions, but then expects the community to do all the work of creating the content. Tom Sawyer wikis are a sure way to fail.

> **What is a Tom Sawyer Wiki**
>
> In the eponymous book, Tom Sawyer, a character invented by Mark Twain, is a clever boy who has to paint a fence as a punishment. Instead of doing the work himself, he pretends to be having a great time doing the work and presents the task to his friends as an honor. Tom Sawyer succeeds in conning his friends out of some of their possessions for the privilege of painting the fence for Tom.
>
> Tom Sawyer wikis are exactly the same thing. Someone creates a wiki, a remarkably easy thing to do nowadays, and then announces to the world that the wiki is open and ready for painting, that is, content creation. The problem is that the wiki has no content. Perhaps the wiki reflects some taxonomy, but there is no there there. The creators of Tom Sawyer wikis expect that other people will now see the task of filling the wiki with content as a great honor.

Lessons from a Book in Progress

The experience of creating the first and second versions of the BPX book led us to the realization that we must find a new way. The Communication by Design process worked well and a strong first draft was improved by review both from the content owners, subject matter experts, and from the community. The second draft is even stronger and reflects many new ideas.

However, the content creation process did not draw as many comments and contributors as expected from the BPX community. While there could be many reasons for this, we suspect that even though the writing team went first, the BPX book project still seemed to many like a Tom Sawyer project in an important way. Community members were not asked to create the first draft, and because of this they did not feel a strong connection to the first draft either.

The question is how can the community feel genuine ownership but not be burdened with having to do all of the work?

How to Increase Community Participation

The new method proposed here will have to be tried out on a new project, or on additional chapters for this project. If this process works well, it may represent a significant step forward in harnessing the power of a community authoring.

To avoid a Tom Sawyer wiki, someone must go first. Great content comes when a properly prepared individual, with experience and writing skill, absorbs a body of knowledge and then creates a narrative to communicate that knowledge to others. A properly prepared individual will always beat a self-organizing community in the amount and quality of content that is created.

But how can the community play a role?

The suggested remedy to the challenge of community ownership versus individual virtuosity is embedded in the following method for community content creation that is accelerated by a professional team.

For this to work, the community must be brought in much earlier, participating in the design process and having a say about the mission of the content. Then they can feel that they went first in creating the content. All of this interaction would take place on a wiki.

Step 1: Exploration: Mapping questions and concepts
- In this step, the community is asked what content should be created. The community is requested to supply questions that need to be answered, topics that need to be explained, cases that should be made, research materials that would be helpful, and so on.
- At the same time, community members will be asked about the role they are willing to play. Are they willing to be interviewed? To refer experts to the project? To review content? To write?

Step 2: Determine the mission
- Based on the ideas gathered during the exploration, at least two and possibly more missions for the content are created. These missions

must contain a statement of what content would be created, the sort of subject matter experts who would be involved, and a listing of the community members who would support the mission based on their willingness to participate in the ways listed above.
- The missions would be open for comment and suggestion and refinement.
- Eventually the community would vote on which mission to pursue.

Step 3: Research Coverage
- The writing team would then take the mission that was chosen and execute the Communication by Design process as described above, using the community members and other participants suggested by the content owners.
- A content design would be created and commented on. Research into all topics would be performed. A first draft would be created and commented on by the community and other reviewers.

Step 4: Refactoring and Filling Holes
- Based on the review of the first draft, the content design may be improved, as it was in the BPX book from the first to the second version, and further research may be performed. The comments on the first draft and, if needed, additional research and interviews would be used to create a second draft.

Step 5: Refinement and Polishing
- The second and subsequent drafts would be created and reviewed until the content was declared complete.

Using this process, the community would have ownership and a more substantial role, but would not have to go first and fill in the blanks in a Tom Sawyer wiki.

Will this work? Perhaps; perhaps not. The only way to know is to try. We hope the BPX community can start such a project in the near future. We hope you have enjoyed this first stab at defining the business process expert role and we hope that it clarifies many of the open questions. We look forward to more input as the community grows. On behalf of the community, I would like to thank you for your current and future contributions.

Marco ten Vaanholt

To order copies of this book:

To order fewer than 10 copies:

Please visit *www.evolvedtechnologist.com/bpx*
or order directly from Amazon:

Go to *www.Amazon.com*
and search *"process first bpx"*

To order 10 or more copies:

Evolved Media offers a bulk discount for 10 or more copies.

To order, contact **Ruth Voorhies** at:

books@evolvedmedia.com
(646) 827-2196

Printed in the United States
143437LV00002B/87/P